Cambridge Elementary Classics

VERRES IN SICILY

VERRES IN SICILY

Selections from

CICERO'S VERRINE ORATIONS

Compiled and edited by

H. GROSE-HODGE, M.A.

and

E. W. DAVIES, M.A.

CAMBRIDGE

AT THE UNIVERSITY PRESS

1935

Published by the Syndics of the Cambridge University Press
Bentley House, 200 Euston Road, London NW1 2DB
American Branch: 32 East 57th Street, New York, N.Y.10022

ISBN: 0 521 04653 X

First edition 1935
Reprinted 1939 1941 1945
1947 1950 1955 1960
1961 1964 1971 1973

Printed in Great Britain
at the University Printing House, Cambridge
(Brooke Crutchley, University Printer)

CONTENTS

PREFACE

The production of this little volume is prompted by the belief, based on experience, that boys will enjoy reading Latin literature provided that it is concerned with something that interests them; and that not all Latin literature, certainly not all Cicero, can reasonably be so described. The experiment of offering them something on the lines of their natural tastes was recently made in the selection entitled "Murder at Larinum", compiled from the narrative portions of Cicero's speech *pro Cluentio*. Its success seems to warrant this companion volume.

Here, then, is a selection from the orations against Verres. Once again we have a picturesque villain, a series of incidents full of human interest, and an instructive picture—highly coloured perhaps, but not substantially untrue—of life as it was lived in the last days of the Republic, this time in the provinces. Once again innocence is triumphantly defended and guilt exposed, a result less common in the annals of the provinces than in those of Rome, for Verres as a governor was unusual only in his thoroughness. The figure of Verres, unscrupulous but vacillating, rapacious but without discrimination, devoid of human feelings but passionately fond of roses, is supported by a number of minor characters, sketched with a few strokes by the master hand of Cicero. The governor's staff, the rascally agents whom he kept to tell him what to steal, his soothsayer and his orchestra; and on the other hand the worthy provincial notables, the Romans who did business with them, the rich foreign tourists—all these appear, not as mere names in a classical work, but as human beings, living and suffering, who move us to indignation, sympathy and not least to mirth.

It is probable that on no speech did Cicero ever bestow the pains that he gave to the Verrines. The result is a simplicity, an apparent artlessness, which, while it masks the years of practice, the months of special preparation, makes for easy and pleasant reading. The vocabulary is wide without being too difficult, the style lucid and free from ponderous elaboration. It is believed that boys in the middle school who read this story will improve not only their understanding of Roman life and character, but their knowledge of both Latin and English.

Some account of the setting of the story is given in the Introduction. Boys are strongly advised not to treat it with the contempt which they tend to suppose that all Introductions deserve: it will make their reading of the text both easier and more profitable. All boys, on the other hand, read Notes—in the hope of saving themselves trouble. But such is not the object with which these Notes have been compiled, and if consulted with this motive alone they may cause disappointment. Those, however, who take the trouble to think things out for themselves will find their thinking directed and rewarded by an intelligent use of the notes.

The Vocabulary is frankly a labour-saving device. Boys would learn much more from the use of a full-sized dictionary, but their preparation of the Text would take them much longer. They should at all events appreciate the limitations of this and of all other Vocabularies, and should realize that it does not propose to offer them the exact and invariable English equivalent for each Latin word. No such equivalent in fact exists in any language for any word in any other language. It is hoped that the reading of this volume will increase their literary sense and their ear for Latin, by which alone they will be enabled to decide which of many English words is the best to choose as the nearest approximation to the Latin word in its particular context.

Our grateful thanks are due for scholarly criticism and advice to our friend and colleague, Mr C. M. E. Seaman.

H. G.-H.
E. W. D.

Bedford
Feb. 1935

INTRODUCTION

Roman Provincial Government

It has been said that the British acquired their Empire "in a fit of absence of mind". Not so the Romans. They, at all events, were reluctant to add to their responsibilities by annexing territories overseas. But "Empires have their natural frontiers, and until these were reached, no lasting halt was possible". Add to this the inevitable tendency of organization to dominate the unorganized, once contact has been established, and we need seek no further for the causes that underlay the growth of the Roman and most other Empires. When Sicily, the first province, was annexed, Rome was still a small Republic, little more than a city-state. Three and a half centuries later, the emperor Trajan annexed Mesopotamia, the last province to be acquired, and so completed the Roman sovereignty over the world.

Whoever may have regretted the change from Republic to Empire, it is certain that the provincials did not. "A democracy", said the Athenian demagogue, Cleon, "is incapable of ruling an empire"; and the Roman oligarchy proved even more so. As a provincial governor, Verres may have been more than ordinarily bad, but there is plenty of evidence that he was but the average "writ large". Cicero describes the conduct of his predecessor in

Cilicia as that not of a man, but of some wild beast; and what is no less significant is the blindness of a well-meaning governor, like Cicero himself, to what would now be considered some of his most elementary obligations. To explain the career of a Verres we must look not so much at the defects of his personal character as at those of the system which made him possible.

The particular feature of the Roman system which was perhaps more responsible than any other for the misgovernment of the provinces was the fact that the ruling caste was not free to engage in trade. The aristocracy consisted, at the beginning of the first century B.C., no longer of the old patrician families but of those who by their energy, brains or lack of scruple, had succeeded in occupying the leading positions in the State. In other words, it was an aristocracy of office, which tended, however, to become fixed and hereditary. Thus, as soon as ability made its way to the top, it found enterprise barred to it save in one direction—that of the twin careers of politics and arms. Now success in politics, of which the attainment of the consulship was the chief, depended on a series of popular elections to this and to subsidiary offices; and the people, who thus had power without responsibility, expected to be paid for their votes at a progressively increasing rate. The vast sums which the ambitious politician had to spend in this way he could not make and must therefore borrow. Julius Caesar, for instance, who started his political career at the time when Verres was in Sicily, was in debt to the tune of 25,000,000 sesterces within ten years, though he had not yet been elected consul.

Debts thus accumulated must be repaid—but how? Only one way was open to the successful, but ruined, politician, the exploitation of some province as its governor. This, therefore, was the point of view from which he looked forward to the term of provincial administration which, as a propraetor or proconsul, he might expect. He might hope, as its military commander, to win for himself power and reputation as well; but be that as it might, he must at least pay his debts.

In this ambition he could hardly fail, for within his province his will was law. There was, as Cicero wrote to his brother Quintus, "no appeal, no means of complaint, no Senate, no public meeting". The whole power of the state, executive, judicial and administrative, was his and his alone. He did not have to look for the money; it fell into his lap. As supreme judge, he could put justice up for sale at his own price. As head of the government, he need only stay where he was and his subjects vied with each other in securing, for adequate consideration, his good will. As commander-in-chief he could billet his soldiers where he would and be well paid by those who were reluctant to entertain them. We are told it made little difference to a provincial city whether it was captured by a hostile army or rescued by a Roman.

It must be remembered, too, that a Roman governor had none of those restraints of tradition, education or religion which we take for granted. Born and brought up in a slave-owning community, he early learned that not all human beings had human rights. He grew up imbued with a sense of his own superiority as a Roman over all "lesser breeds without the law".[1] There was no public opinion in favour of just dealing; nothing in the Roman religion, nothing in the accepted code of morals, which cared for the interests of those who were not Romans. All conquered territory was regarded as passing into the ownership of the Roman people, the original owners remaining there on sufferance. Government in the interest of the governed was an idea that had not yet been born.

Even if a particular governor did not care, or did not need, to oppress the provincials, it was difficult for him to prevent others doing it. His staff, in any case, expected opportunities of plunder proportionate to his rank as well as theirs. But difficult as it was for him to restrain them, there were others whom he could not restrain—the *publicani*. They alone in his province were not responsible to him, but to those to whom he was himself responsible, the Roman people. Formed into immense *societates* (joint stock companies) in which the general public invested largely, their sole concern was to extract from the province, whose taxes they had contracted

[1] The *ius civile*, Roman Citizen Law, was not applicable to provincials.

to collect, the largest possible amount. For they had to make, not only the sum they had paid for their contract, and sufficient to cover the cost of collecting it, but enough to pay a handsome dividend to their shareholders. A bad governor was only too ready to co-operate with them for the enrichment of them both. Against a more scrupulous governor they always had a weapon in reserve, the threat of prosecution.

In fact, prosecution by one party or another was the lot which all governors, whether good or bad, might have to expect on returning to Rome at the end of their term of office; but the good were probably in greater danger of conviction.

A bad governor might be prosecuted by the provincials. A series of statutes against extortion (*de repetundis*) and the ease with which they were evaded proves both the need and the difficulty of checking the abuses which went on despite them. Even if the provincials were left with sufficient spirit and optimism to prosecute, they had but little chance of gaining a verdict from the Roman courts. The right to sit as jurors, once confined to the patricians, was disputed by the new monied class, the *equites*, in the course of an embittered struggle which went on, with varying result, between the reforms of Gaius Gracchus in 123 B.C. and those of Pompeius and Crassus in 70 B.C. But whatever the constitution of the courts at a given moment, a patrician jury would look indulgently on sins of which they had been, or hoped to be, guilty themselves; and the equestrian order (to which the *publicani* belonged) could be trusted to condone a bad governor's extortion, provided that he had connived at their own. If a good governor had refused his connivance, he might know what to expect.

So, between the governor and the tax-collector, the hapless provincial was ground as between the upper and the nether millstone. "Words cannot express", said Cicero, and he said it in the Forum at Rome, "the bitterness with which we are hated among foreign nations, owing to the wanton and outrageous conduct of the men whom we have sent to govern them."[1] Small wonder that he added privately, "they are absolutely sick of life".[2]

But in the issue of any trial there must be a certain element of doubt; and the retiring governor had only one way of removing it—by bribing the court. This method, though generally certain, was expensive; and a large sum must be allowed for it in considering the total profit which a governor must make out of his province. And so it came about that the ordinary proconsul went to his province with the confident expectation of making not one, but three fortunes: one to pay his debts, one to bribe his judges and one to keep for himself. Verres congratulated himself on having fully realized this expectation.

Only a revolution could effect the reform of this vicious system, in the maintenance of which every

[1] Cic., *pro lege Manilia*, 22. 65.
[2] *Taedet omnino vitae.*

class of the Roman people was interested. The revolution came; and Cicero, who knew better than any how badly it was needed, sacrificed his life in a futile attempt to resist it. It was left to successive Caesars to restore to the provinces some of the liberties which they took from Rome; and to build up an Imperial Civil Service which was without equal till the nineteenth century, and which gave to the Roman Empire a degree of happiness and prosperity unknown to the world before and perhaps since.

SICILY

The island of Sicily is geographically connected both with Europe and Africa; and in prehistoric times, when the eastern and western portions of the Mediterranean were lakes, it formed a land-bridge between them. Its mountains are but a continuation of the Apennines in the north-east, hardly interrupted by the strait of Messina, and are themselves continued by the Atlas Mountains to the south-west, the whole range once forming an unbroken "fold" in the earth's surface. Its earliest inhabitants were of Italian origin; but Sicilian history, like that of Italy itself, really opens with the coming of the Greeks.

The first Greek colony was founded, according to tradition, in 735 B.C., followed by Syracuse a year later; and by the end of the sixth century B.C., despite Phoenician settlements planted from Africa upon the western sea-board, Sicily was predominantly Greek. The island changed hands more than once in the course of the next four centuries, but its civilization remained Greek throughout and has bequeathed its lovely relics, through twenty-five centuries of catastrophes both natural and political, to be the admiration of to-day. It may be presumed, without being regretted, that their marble will outlast our ferro-concrete.

No native power has ever ruled Sicily; but, lying between Carthage on the one side and Greece and Rome on the other, the island played an important part in the history of all three countries and was, indeed, the cock-pit of the Mediterranean world. Instigated by Persia in 480 B.C. Carthage launched a great attack on the Greek colonies there, which was defeated by Gelon of Syracuse on the same day (so Herodotus was told) that the Persians were themselves defeated by Athens at the battle of Salamis. In the harbour of Syracuse there was sunk in 413 B.C., with the Athenian Armada, the Athenian hope of an empire in the west and the chance of winning the Peloponnesian War. It was Syracuse, too, which alone held out when, in the century of disruption and weakness which followed the victory of Sparta, Carthage took the opportunity to dominate the rest of the island. The hopes of Greek sovereignty were revived by Pyrrhus in 279 B.C., but this time it was Rome that put an end to them for ever.

Rome and Carthage, who had entered into a short-lived alliance in order to deal with Pyrrhus, were now left to fight out in Sicily the age-long conflict between Aryan and Semite. The First Punic War broke out there in 265 B.C., and, by the terms of the peace which concluded it in 241 B.C.,[1] Sicily, ceded by Carthage, was annexed as the first Roman province. Twenty years later a last attempt by Carthage to recover the island developed into the Second Punic War, whose issue left Rome mistress of Sicily and soon to be mistress of the world.

Syracuse threw in her lot with Carthage and in 214 B.C. shut her gates in the face of the consul Marcellus, who now settled down to a siege. For two years the city walls and the skill of the famous engineer Archimedes[2] kept him out, but in 212 B.C. a traitor showed him the way in and the Romans became masters of Syracuse. Marcellus spared the lives of the inhabitants—though Archimedes was killed in the confusion—and refrained from destroying the public buildings, but he gave the city up to unrestrained pillage. So completely was it looted that many Syracusans, unable to keep themselves alive, had no alternative to selling themselves into slavery. And yet, such was the insensibility of the Romans and such the standard of conduct expected of a victorious general, that the clemency and liberality of Marcellus have been extolled by many ancient writers, including Cicero, who contrasts these qualities as displayed by the conqueror

[1] The eastern end of the island was not included till after the fall of Syracuse in 212 B.C.

[2] Cicero discovered the tomb of Archimedes while serving as quaestor in Sicily in 75 B.C.

of Syracuse with the unscrupulous rapacity of its later governor, Verres. Actually it was Marcellus who set the example. From the looted city he carried off an immense booty which included, besides the money in the treasury, many of the works of art which adorned the public buildings and which were now to grace his triumph and the temples of Rome. He thus did more, perhaps, than any one man to stimulate among his countrymen that interest in Greek art which became increasingly the fashion at Rome.

For a thousand years after the Second Punic War Sicily remained under the *imperium* of Rome; [1] but there is evidence of commercial relationships between the two long before there was any political connexion, and at a time when Rome appears to have enjoyed them with no other country but Greece. The Sicilian currency, weights and measures were, as we should say, linked with those of Rome; and one of the few dated facts in Roman commercial history is the arrival from Sicily of the first barber to ply his trade in Italy.

But Sicily's chief importance to Rome was as a corn-producing country. The broad and fertile belt of coast-land that surrounds the comparatively barren interior grew enough corn to relieve the shortage that threatened the population of Rome after the disaster of Cannae in 216 B.C. and again

[1] Except for an interval in the fifth century when it was seized by Genseric the Vandal. It was recovered from the Goths by Belisarius in the sixth century.

during the Social War of 90–88 B.C. By the end of the second century Sicily was known as one of the three granaries of Rome, the other two being Egypt and Sardinia. The increase in the supply of corn from these last-mentioned countries probably accounts for the decline in the importance of Sicily under the Empire.

Such was the worth, and such had been the cost, of Sicily to Rome that the government of the Roman Republic looked on this, the largest and the fairest island in the Mediterranean, with special jealousy and treated the Sicilians with special consideration. Governed as a single province under a praetor or propraetor, it always had two quaestors, of whom Cicero had been one in 75 B.C. Of its sixty towns which enjoyed municipal rights, three were "allied cities" (*civitates foederatae*) and so nominally independent. Five more (*civitates immunes et liberae*) were exempt from all taxation and from the ordinary jurisdiction of Roman magistrates. The rest retained their own magistrates and territories, subject to the payment of a tenth of their produce to Rome. These tithes, which were habitually farmed out, were paid in kind and were administered according to the principles originally laid down by Hiero II, king of Syracuse, in the third century. Under the emperor Augustus, Sicily became a senatorial province, governed by a proconsul, and was ultimately regarded as an integral part of Italy.

Verres and Cicero

In the year 73 B.C., when our narrative opens, Sicily was the acknowledged queen of the Roman provinces. Besides the importance of its corn-supply, the nearness of the island to Italy and its ancient connexion with Rome, to say nothing of its climate and scenery, made it a favourite resort of Roman capitalists, who lived on terms of friendly intercourse with the families of wealthy and cultured Sicilians. The governorship of this delectable island was a "plum" indeed; and in this year it fell by lot to Gaius Verres. No one, we may be confident, would have been more appreciative of his good fortune.

Unfortunately for us—and still more so for him—we know almost nothing about Verres except what Cicero has chosen to tell us. His father was a senator—probably raised to that position from humbler rank through the favour of Sulla. He himself was quite an ordinary person if judged, not by our standards, but by those of his time and class. He behaved in Sicily as he had already behaved elsewhere and as most of his contemporaries would have liked to behave anywhere. He differed from others in the greater opportunities that he had and the great thoroughness with which he used them. To say that he was a man of brutal passions and that he gratified them without scruple; that he was lazy, greedy and cruel; and that he surrounded himself with a following of the same kidney, amounts to little more than to saying that he belonged to an incurably corrupt class of men—the provincial governors in the declining years of the Roman Republic. His misconduct of his office differed from that of others in degree rather than in kind. His trial therefore derives its interest from the cause rather than the criminal.

The praetorship of Verres, extended to nearly three years by the troublous times which kept his successor in Italy, coincided with a critical period in the fortunes of Sicily. Thirty years before his arrival, the second of two formidable slave-revolts had inflicted damage so severe that it is doubtful if, despite its great natural powers of recovery, the prosperity of Sicily at this time was as great as Cicero makes out, though the milder rule of his predecessor, Sacerdos, had done much to restore it. But if the interior was now safe, the coast, always the most productive and important part of the island, was liable to attack at any moment by the dreaded pirates who at this time swarmed over the Mediterranean.

These pirates called themselves Cilicians, after the region whose mountains and indented coast-line provided the best base for their operations; but they consisted in fact of the ruined and the desperate of all nations, whom the incessant wars of the Romans and those who sought to resist them had deprived of all other hopes of livelihood. United in a military-political organization, with their own code of law and even of honour, they made it their boast to live at war with all the world; and in the Mediterranean world no sea-port and not all inland cities were safe from the sudden descent of their small, swift vessels. Such sea-borne trade as was not brought to a standstill must first pay toll to the pirates—countless sea-ports paid them a regular tribute. Even Roman troop-ships had to wait for rough weather before they dared put to sea.

The enemies of Rome found in the pirates a valuable ally. It was their alliance with Sertorius in 79 B.C. which induced the Senate to send the powerful expedition which drove them from Cilicia, only to find a fresh base in Crete. Another well-equipped force sent against them in 71 B.C. was a costly failure; and it was left to the great Pompeius, armed with supreme power and backed by the whole resources of the state, to make a final settlement with the pirates in 67 B.C.—and he had to make it as much by conciliation as by force of arms.

As praetor, Verres found himself responsible for the prosperity of his province within and its security without, at a time when both were matters of special concern to Rome. Spartacus and his army of gladiators and slaves were ravaging the heart of Italy. The rebel Sertorius was master of Spain and was in close alliance with the great Mithridates in Asia Minor. The pirates, whose power in the Mediterranean was now unchecked and almost unopposed, were in league with both. If ever the corn-supply of Sicily was needed in Rome, it was needed now, and the task of making it secure called for unusual qualities of vigilance, determination and integrity on the part of its governor. How far our hero rose to the occasion the narrative will disclose. Suffice it now to say that, great as was the havoc wrought in Sicily by the slave war of 103 B.C., the loss sustained through the praetorship of Verres was hardly less.[1]

Slight interest was ordinarily taken at Rome in scandals of provincial mismanagement; but Verres had gone altogether too far and at the wrong time. Not content with his natural prey, the Sicilians, he had treated the Roman citizens in his province with no less cruelty and contempt, till even the Senate began to show a certain restiveness and Verres' own father wrote to Sicily begging him to be more careful. The parental advice (as so often) came too late; and when at last Verres left his province, all the hate and resentment, repressed for three long years, burst out in a chorus of denunciation and a clamour for vengeance. It reached the ears of the Roman people at a time when the military failures and the judicial corruption of the Senate made them more disposed to make an example of a senatorial governor and to watch with special jealousy the conduct of a senatorial court.[2] It was to the fear,

[1] "In four of the most fertile districts of Sicily 59 per cent. of the landholders preferred to let their fields lie fallow than to cultivate them under this régime." Mommsen, vol. IV, p. 531.

[2] The senatorial monopoly of the right to serve as a juror was abolished in the year of Verres' trial.

rather than to the justice, of the court that the prosecution made its first appeal.

Six years earlier Cicero had attained the age of thirty, and was therefore eligible to stand for the first office in the *cursus honorum*, that of quaestor. That he was elected, without the advantage of patrician birth or the support of a powerful interest, is the best evidence of his popularity and his reputation at the bar. The lot decided that he should serve in Sicily; and during the year 75 B.C. he contrived, by a combination of tact and strict personal integrity, to increase the export of corn to Rome and at the same time to secure the gratitude and good will of the Sicilians. Always elated by success and ready to accept flattery at its face-value, it was a shock to him, on returning to Rome, to find the capital apparently unaware of his exploits. But the lesson was a salutary one. He laughed at his own disappointment and resolved to base his claim to popular favour upon the securer foundation of hard work at his profession.

To this, then, he devoted the next four years. Of his practice in the courts during that time we know almost nothing, but we have abundant proof both of his efforts and of his success in the maturity of the power, as an orator and as a lawyer, with which he emerged from them. When the year 70 B.C. opened, he stood in the foremost rank of Roman advocates. At its close he was to stand alone in unchallenged supremacy.

Fate and his own merit thus pointed to Cicero as

the man to whom the distressed Sicilians should entrust their cause. He had personal knowledge of Sicily and they had personal knowledge of him. Theirs was not at first sight the kind of case he would have chosen—he preferred always to speak for the defence—but he yielded to the entreaties of his friends and, once committed, attacked the case with a degree of energy and determination that confounded the tactics of his opponents. Verres, backed by the Metelli and other powerful families, and never hampered by scruple, prepared by every means to defend his ill-gotten gains. As his counsel, he briefed Quintus Hortensius, the consul-elect, whose forensic reputation was challenged only by Cicero. The two rivals were now face to face, and Hortensius at least knew that it would be a duel to the death.

From the start, every obstacle was put in Cicero's way. An attempt was made to take the case out of his hands altogether; but in vain. Cicero demanded, and was granted, 110 days in which to prepare his case, and immediately started for Sicily. In less than half the time allotted he had toured from end to end of the island, sifting evidence, taking depositions, and collecting witnesses. With these he returned to Rome, nearly two months before either his friends or his enemies expected him, only to find that Hortensius was now determined to obstruct the trial till the following year, when he himself would be consul and the court might prove more open to influence or to bribery. Once again Cicero was too clever for him. He opened his case very briefly on August the 5th and proceeded at once to lay before the court the vast mass of evidence, both oral and documentary, which he had brought with him—formidable in its bulk and irresistible in its conclusions. His case was overwhelming and Verres knew it. Without giving Hortensius the chance to say a word in his defence, he fled from Rome and went into voluntary exile.

But Cicero was not the man to leave in obscurity the great speech which he was prepared to deliver should the trial run its normal course. He therefore published what we now know as the Verrine Orations, and it is from them that the contents of this volume have been selected. The selections represent the merest fragment of the imposing whole; but they should suffice to illustrate the range and brilliance of Cicero's gifts—his amazing command of language; his humour and his sarcasm, his devastating power of invective; his ability to paint a word-picture; the way in which he avoids monotony by changing his style from the simplest and barest narrative to the most impassioned rhetoric. To all this must be added his ability as an artist to identify himself with his part and to feel the emotions which he wished his audience to feel. But only the entire speech can really bring home another gift without which all his brilliance would have been of little use—his almost incredible capacity for work and the command of detail which resulted.

These qualities, matured by years of practice and the severest self-discipline, he brought to all his speeches. But we may credit his denunciation of Verres with another quality for which we look elsewhere in vain—nothing less than a passionate sincerity. When he spoke of the devastation of Sicily, he saw her once fair cities and prosperous countryside before his eyes. When he inveighed against the tyranny of the governor, he spoke with the personal feeling of one who had proved himself, and was to prove himself again, a just judge and a humane administrator. It is true that in the person of Verres Cicero the equestrian was attacking the creature of a corrupt and exclusive aristocracy, and that the counsel for the defence was his old and only rival. But Cicero was pleading before all the cause of decency, honest dealing and clean living; and he cared truly and deeply for these things.

It is somehow sad, as well as difficult, to remember, as we read the orator's burning words, that they never rang out in court to lash the craven Verres to despair and his judges to virtuous indignation. It is sadder still to reflect that if "he laughs loudest who laughs last", the smile was (if we may adapt another quotation) "on the face of the tiger". Verres, as we have said, anticipated the sentence of the court and retired to exile at Marseilles; and there we may suppose that when Milo joined him later on they discussed together, over a dish of the famous mullets, the man who had not needed to accuse the one or dared to defend the other.[1] Verres lived in exile for twenty-seven years in open enjoyment of so large a part of the spoils of his province as to make him indifferent to public opinion; and doubtless he watched with unholy glee the troubles of the country that had expelled him and the ruin of the judges who had tried him. Sweetest of all must have been the news of the murder in 43 B.C. of his old enemy Cicero by the only man who had even better cause than himself to hate him—Mark Antony. The victim of the Philippics had avenged the victim of the Verrines.

But the satisfaction of Verres was short lived. *Raro antecedentem scelestum deseruit pede Poena claudo.*[2] Poetic justice overtook him at last, and the treasures that he had looted from Sicily cost him his life; for Antony was no less covetous than he was vengeful. Verres was proscribed as well as Cicero and survived his accuser by a few weeks only.[3]

[1] Cicero was briefed in 52 B.C. to defend Milo on a charge of violence; but his courage failed him at the sight of Pompey's soldiers lining the Forum and Milo, like Verres, went into exile at Marseilles. Cicero once again published the speech he had meant to deliver (the *pro Milone*) and sent a copy of it to Milo, who replied that he was glad it had never been delivered as "I must have been acquitted and then I should never have known the flavour of these Massilian mullets".

[2] "Rarely does vengeance, albeit of halting gait, fail to o'ertake the guilty though he gain the start." Horace, *Odes*, III. 2. 31, tr. E. G. Bennett, Loeb Library.

[3] Whether Verres or Cicero actually perished first is not strictly known. The view expressed here is that of Long.

Prooemium

§ 1

I must begin by mentioning the special claims which the Sicilians have upon Rome; and shall be content if you show Verres just so much consideration as he showed them.

Multa mihi necessario, iudices, praetermittenda sunt, ut possim aliquo modo aliquando de eis rebus quae meae fidei commissae sunt dicere: recepi enim causam Siciliae: ea me ad hoc negotium provincia attraxit. 5

Atque antequam de incommodis Siciliae dico, pauca mihi videntur esse de provinciae dignitate, vetustate, utilitate dicenda; nam cum omnium sociorum provinciarumque rationem diligenter habere debetis, tum praecipue Siciliae, iudices, pluri- 10 mis iustissimisque de causis: primum, quod omnium nationum exterarum princeps Sicilia se ad amicitiam fidemque populi Romani applicavit: prima omnium provincia est appellata, prima docuit maiores nostros quam praeclarum esset exteris 15 gentibus imperare; sola fuit ea fide benevolentiaque erga populum Romanum ut civitates eius insulae, quae semel in amicitiam nostram venissent, numquam postea deficerent, pleraeque autem et maxime illustres in amicitia perpetuo manerent. Neque tam 20 facile opes Karthaginis tantae concidissent, nisi illud et rei frumentariae subsidium et receptaculum classibus nostris pateret.

Videor mihi gratum fecisse Siculis, quod eorum
25 iniurias meo labore, inimicitiis, periculo sum persecutus; non minus hoc gratum me nostris civibus intellego fecisse, qui hoc existimant, iuris, libertatis, rerum fortunarumque suarum salutem in istius damnatione consistere. Quapropter de istius prae-
30 tura Siciliensi non recuso quin ita me audiatis ut, si cuiquam generi hominum sive Siculorum sive nostrorum civium, si cuiquam ordini sive aratorum sive pecuariorum sive mercatorum probatus sit, si non horum omnium communis hostis praedoque
35 fuerit, si cuiquam denique ulla in re unquam temperaverit, ut vos quoque ei temperetis.

§ 2

The unhappy province knew what to expect from his government and has hardly yet recovered from it; for as governor he showed no respect for justice, duty, life or property.

Qui, simul atque ei sorte provincia Sicilia obvenit, statim Romae, antequam proficisceretur, quaerere ipse secum et agitare cum suis coepit quibusnam rebus in ea provincia maximam uno anno pecuniam
5 facere posset. Nolebat in agendo discere, sed paratus ad praedam meditatusque venire cupiebat. O praeclare coniectum in illam provinciam omen communis famae atque sermonis, cum ex nomine
10 istius quid iste in provincia facturus esset perridicule homines augurabantur! Etenim quis dubitare posset, cum istius in quaestura fugam et furtum recognosceret, cum in legatione oppidorum fano-

rumque spoliationes cogitaret, cum videret in foro latrocinia praeturae, qualis iste in quarto actu improbitatis futurus esset? 15

Iam vero omnium vitiorum suorum plurima et maxima constituit monumenta et indicia in provincia Sicilia, quam iste per triennium ita vexavit ac perdidit ut ea restitui in antiquum statum nullo modo possit, vix autem per multos annos inno- 20 centesque praetores aliqua ex parte recreari aliquando posse videatur. Hoc praetore Siculi neque suas leges neque nostra senatus consulta neque communia iura tenuerunt: tantum quisque habet in Sicilia quantum hominis avarissimi et 25 libidinosissimi aut imprudentiam subterfugit aut satietati superfuit. Nulla res per triennium nisi ad nutum istius iudicata est, nulla res tam patria cuiusquam atque avita fuit quae non ab eo imperio istius abiudicaretur. Innumerabiles pecuniae ex 30 aratorum bonis novo nefarioque instituto coactae, socii fidelissimi in hostium numero existimati, cives Romani servilem in modum cruciati et necati, homines nocentissimi propter pecunias iudicio liberati, honestissimi atque integerrimi absentes 35 rei facti indicta causa damnati et eiecti, portus munitissimi, maximae tutissimaeque urbes piratis praedonibusque patefactae, nautae militesque Siculorum, socii nostri atque amici, fame necati, classes optimae atque opportunissimae cum magna igno- 40 minia populi Romani amissae et perditae. Idem iste praetor monumenta antiquissima partim regum locupletissimorum, quae illi ornamento urbibus

esse voluerunt, partim etiam nostrorum impera-
45 torum, quae victores civitatibus Siculis aut dede-
runt aut reddiderunt, spoliavit nudavitque omnia.
"At enim haec ita commissa sunt ab isto ut non
cognita sint ab omnibus." Hominem esse arbitror
neminem, qui nomen istius audierit, quin facta
50 quoque eius nefaria commemorare possit, ut mihi
magis timendum sit ne multa crimina praeter-
mittere quam ne qua in istum fingere existimer.
Neque enim mihi videtur haec multitudo, quae ad
audiendum convenit, cognoscere ex me causam
55 voluisse, sed ea quae scit mecum recognoscere.

PART I

VERRES THE JUDGE

The Story of Sopater

§ 3

Sopater, accused on a charge of which he had already been acquitted, was forced to bribe Verres if he wanted a verdict in his favour.

Iam vero in rerum capitalium quaestionibus quid ego unam quamque rem colligam et causam? Ex multis similibus ea sumam, quae maxime improbitate excellere videbuntur. Sopater quidam fuit Halicyensis, homo domi suae cum primis locuples 5 atque honestus; is ab inimicis suis apud C. Sacerdotem praetorem rei capitalis cum accusatus esset, facile eo iudicio est liberatus. Huic eidem Sopatro idem inimici ad C. Verrem, cum is Sacerdoti successisset, eiusdem rei nomen detulerunt. Res 10 Sopatro facilis videbatur, et quod erat innocens et quod Sacerdotis iudicium improbare istum ausurum non arbitrabatur. Citatur reus; causa agitur Syracusis; crimina tractantur ab accusatore ea, quae erant antea non solum defensione, verum 15 etiam iudicio dissoluta. Causam Sopatri defendebat Q. Minucius, eques Romanus in primis splendidus atque honestus vobisque, iudices, non ignotus. Nihil erat in causa quod metuendum aut omnino quod dubitandum videretur. Interea istius libertus 20 Timarchides, qui est rerum huiusce modi omnium

transactor et administer, ad Sopatrum venit: monet hominem, ne nimis iudicio Sacerdotis et causae suae confidat; accusatores inimicosque eius
25 habere in animo pecuniam praetori dare; praetorem tamen ob salutem malle accipere, et simul malle, si fieri posset, rem iudicatam non rescindere. Sopater, cum hoc illi improvisum atque inopinatum accidisset, commotus est sane, neque in praesentia
30 Timarchidi quid responderet habuit, nisi se consideraturum quid sibi esset faciendum; et simul ostendit se in summa difficultate esse nummaria. Post ad amicos rettulit; qui cum ei fuissent auctores redimendae salutis, ad Timarchidem venit. Expo-
35 sitis suis difficultatibus hominem ad HS LXXX perducit, eamque ei pecuniam numerat.

§ 4

Sopater was next told that he had not paid enough, but he came into court relying on the honesty of the jury.

Posteaquam ad causam dicendam ventum est, tum vero sine metu, sine cura omnes erant, qui Sopatrum defendebant: crimen nullum erat; res erat iudicata; Verres nummos acceperat: quis posset dubitare
5 quidnam esset futurum? Res illo die non peroratur, iudicium dimittitur. Iterum ad Sopatrum Timarchides venit, ait accusatores eius multo maiorem pecuniam praetori polliceri quam quantam hic dedisset; proinde, si saperet, videret quid sibi esset
10 faciendum. Homo, quamquam erat et Siculus et reus, hoc est et iure iniquo et tempore adverso, ferre

tamen atque audire diutius Timarchidem non potuit. "Facite", inquit, "quod libet; daturus non sum amplius." Idemque hoc amicis eius et defensoribus videbatur, atque eo etiam magis quod iste, 15 quoquo modo se in ea quaestione praebebat, tamen in consilio habebat homines honestos e conventu Syracusano, qui Sacerdoti quoque in consilio fuerant tum cum est idem hic Sopater absolutus. Hoc rationis habebant, facere eos nullo modo posse 20 ut eodem crimine eisdem testibus Sopatrum condemnarent idem homines qui antea absolvissent. Itaque hac una spe ad iudicium venitur.

§ 5

Verres, however, began to dismiss the honest jurors; and finally Sopater's friends, including his counsel, left the court.

Quo posteaquam est ventum, cum in consilium frequentes convenissent idem qui solebant, et hac una spe tota defensio Sopatri niteretur, consilii frequentia et dignitate, et quod erant, ut dixi, idem qui antea Sopatrum eodem illo crimine liberarant, 5 cognoscite hominis apertam ac non modo non ratione, sed ne dissimulatione quidem tectam improbitatem et audaciam. M. Petilium, equitem Romanum, quem habebat in consilio, iubet operam dare, quod rei privatae iudex esset. Petilius recu- 10 sabat, quod suos amicos, quos sibi in consilio esse vellet, ipse Verres retineret in consilio. Iste homo liberalis negat se quemquam retinere eorum qui

Petilio vellent adesse. Itaque discedunt omnes;
15 nam ceteri quoque impetrant ne retineantur; qui
se velle dicebant alterutri eorum qui tum illud
iudicium habebant adesse. Itaque iste solus cum
sua cohorte nequissima relinquitur.

Non dubitabat Minucius, qui Sopatrum defende-
20 bat, quin iste, quoniam consilium dimisisset, illo die
rem illam quaesiturus non esset, cum repente
iubetur dicere. Respondet, "Ad quos?" "Ad me,"
inquit, "si tibi idoneus videor qui de homine Siculo
ac Graeculo iudicem." "Idoneus es," inquit, "sed
25 pervellem adessent ii qui adfuerant antea causam-
que cognorant." "Dic," inquit; "illi adesse non
possunt." "Nam hercule", inquit Minucius, "me
quoque Petilius ut sibi in consilio adessem rogavit",
et simul a subselliis abire coepit. Iste iratus
30 hominem verbis vehementioribus prosequitur,
atque ei gravius etiam minari coepit quod in se
tantum crimen invidiamque conflaret. Minucius,
qui Syracusis sic negotiaretur ut sui iuris dignita-
tisque meminisset, et qui sciret se ita in provincia
35 rem augere oportere ut ne quid de libertate deper-
deret, homini quae visa sunt, et quae tempus illud
tulit et causa, respondit, causam sese dimisso atque
ablegato consilio defensurum negavit. Itaque a
subselliis discessit, idemque hoc praeter Siculos
40 ceteri Sopatri amici advocatique fecerunt.

§ 6

Verres, now thoroughly alarmed, was at a loss what to do, till Timarchides whispered in his ear. Thereupon, after a farcical trial, he pronounced Sopater guilty.

Iste, quamquam est incredibili importunitate et audacia, tamen subito solus destitutus pertimuit et conturbatus est; quid ageret, quo se verteret nesciebat. Si dimisisset eo tempore quaestionem, post, illis adhibitis in consilium, quos ablegaverat, 5
absolutum iri Sopatrum videbat; sin autem hominem miserum atque innocentem ita condemnasset, cum ipse praetor sine consilio, reus autem sine patrono atque advocatis fuisset, iudiciumque C. Sacerdotis rescidisset, invidiam se sustinere tantam 10
non posse arbitrabatur. Itaque aestuabat dubitatione; versabat se in utramque partem non solum mente, verum etiam corpore, ut omnes qui aderant intellegere possent in animo eius metum cum cupiditate pugnare. Erat hominum conventus 15
maximus, summum silentium, summa exspectatio quonam esset eius cupiditas eruptura; crebro se accensus demittebat ad aurem Timarchides. Tum iste aliquando: "Age dic", inquit. Reus orare atque obsecrare, ut cum consilio cognosceret. Tum 20
repente iste testes citari iubet: dicit unus et alter breviter; nihil interrogatur; praeco DIXISSE pronuntiat. Iste, quasi metueret ne Petilius privato illo iudicio transacto aut prolato cum ceteris in consilium reverteretur, ita properans de sella exsilit; 25
hominem innocentem, a C. Sacerdote absolutum,

indicta causa de sententia scribae, medici haruspicisque condemnat.

The Story of Sthenius

§ 7

When visiting Thermae, Verres used to stay with his friend Sthenius, an art-collector, whose treasures he succeeded in taking from him. He next tried to get hold of some famous statues that the town possessed, but Sthenius resisted and foiled him.

Accipite nunc aliud eius facinus nobile et multis locis saepe commemoratum, et eius modi ut in uno omnia maleficia inesse videantur. Sthenius est, hic qui nobis adsidet, Thermitanus, antea multis
5 propter summam virtutem summamque nobilitatem, nunc propter suam calamitatem atque istius insignem iniuriam omnibus notus. Huius hospitio Verres cum esset usus et apud eum Thermis saepenumero fuisset, domo eius omnia abstulit,
10 quae paulo magis animum cuiuspiam aut oculos possent commovere. Etenim Sthenius ab adulescentia paulo studiosius haec compararat, supellectilem ex aere elegantiorem, et Deliacam et Corinthiam, tabulas pictas, etiam argenti bene facti,
15 prout Thermitani hominis facultates ferebant, satis. Quae posteaquam iste omnia abstulit, alia rogando, alia poscendo, alia sumendo, ferebat Sthenius, ut poterat; angebatur animi necessario, quod domum eius exornatam et instructam fere
20 iam iste reddiderat nudam atque inanem; verum

tamen dolorem suum nemini impertiebat; praetoris iniurias tacite, hospitis placide ferendas arbitrabatur. Interea iste cupiditate illa sua nota atque apud omnes pervagata, cum signa quaedam pulcherrima atque antiquissima Thermis in publico 25 posita vidisset, adamavit; a Sthenio petere coepit ut ad ea tollenda operam suam profiteretur seque adiuvaret. Sthenius vero non solum negavit, sed etiam ostendit fieri id nullo modo posse, ut signa antiquissima, monumenta P. Africani, ex oppido 30 Thermitanorum tollerentur.

Erant signa ex aere complura; in his eximia pulchritudine ipsa Himera in muliebrem figuram habitumque formata ex oppidi nomine et fluminis. Erat etiam Stesichori poetae statua senilis incurva 35 cum libro summo, ut putant, artificio facta, qui fuit Himerae, sed et est et fuit tota Graecia summo propter ingenium honore et nomine. Haec iste ad insaniam concupiverat. Etiam, quod paene praeterii, capella quaedam est, ea quidem mire, ut 40 etiam nos qui rudes harum rerum sumus intellegere possumus, scite facta et venuste. Haec cum iste posceret agereturque ea res in senatu, Sthenius vehementissime restitit multaque, ut in primis Siculorum in dicendo copiosus est, commemoravit: 45 urbem relinquere Thermitanis esse honestius quam pati tolli ex urbe monumenta maiorum, spolia hostium, beneficia clarissimi viri, indicia societatis populi Romani atque amicitiae. Commoti animi sunt omnium; repertus est nemo quin mori diceret 50 satius esse. Itaque hoc adhuc oppidum Verres

invenit prope solum in orbe terrarum unde nihil eius modi rerum de publico per vim, nihil occulte, nihil imperio, nihil gratia, nihil pretio posset
55 auferre.

§ 8

Verres' revenge was to arrange for a false charge to be brought against Sthenius, who fled to Rome. He was condemned in his absence, and summoned to return and answer another and graver charge.

Verum hasce eius cupiditates exponam alio loco; nunc ad Sthenium revertar. Iratus et incensus hospitium ei renuntiat; domo eius emigrat atque adeo exit; nam iam ante emigrarat. Eum autem
5 inimicissimi Sthenii domum suam statim invitant, ut animum eius in Sthenium inflammarent ementiendo aliquid et criminando.

Itaque hortari homines coepit ut aliquid Sthenio periculi crearent criminisque confingerent. Dice-
10 bant se illi nihil habere quod dicerent. Tum iste his aperte ostendit et confirmavit eos in Sthenium quicquid vellent, simul atque ad se detulissent, probaturos. Itaque illi non procrastinant: Sthenium statim educunt; aiunt ab eo litteras publicas
15 esse corruptas. Sthenius postulat ut, cum secum sui cives agant de litteris publicis corruptis, senatusque et populus Romanus Thermitanis ita leges suas reddidisset ut cives inter se legibus suis agerent, ut se ad leges reiceret. Iste, homo omnium
20 aequissimus atque a cupiditate remotissimus, se cogniturum esse confirmat; paratum ad causam

dicendam venire hora nona iubet. Non erat obscurum quid homo improbus ac nefarius cogitaret; neque enim ipse satis occultarat: intellectum est id istum agere ut, cum Sthenium sine ullo argumento 25 ac sine teste damnasset, tum homo nefarius de homine nobili atque id aetatis suoque hospite virgis supplicium crudelissime sumeret.

Quod cum esset perspicuum, de amicorum hospitumque suorum sententia Thermis Sthenius 30 Romam profugit: hiemi sese fluctibusque committere maluit quam non istam communem Siculorum tempestatem calamitatemque vitare. Iste homo certus et diligens ad horam nonam praesto est; Sthenium citari iubet: quem posteaquam videt 35 non adesse, dolore ardere atque iracundia furere coepit; Venerios domum Sthenii mittere, equis circum agros eius villasque dimittere. Itaque dum exspectat quidnam sibi certi adferatur, ante horam tertiam noctis de foro non discedit. Postridie mane 40 descendit; Agathinum ad se vocat; iubet eum de litteris publicis in absentem Sthenium dicere. Erat eius modi causa ut ille ne sine adversario quidem apud iniquum iudicem reperire posset quid diceret. Itaque tantum verbo posuit Sacerdote praetore 45 Sthenium litteras publicas corrupisse. Vix ille hoc dixerat cum iste pronuntiat STHENIVM LITTERAS PVBLICAS CORRVPISSE VIDERI; et hoc praeterea addit novo modo nullo exemplo, OB EAM REM HS D VENERI ERYCINAE DE STHENII BONIS SE 50 EXACTVRVM, bonaque eius statim coepit vendere; et vendidisset, si tantulum morae fuisset quo minus

ei pecunia illa numeraretur. Ea posteaquam numerata est, contentus hac iniquitate iste non fuit;
55 palam de sella ac tribunali pronuntiat, SI QVIS ABSENTEM STHENIVM REI CAPITALIS REVM FACERE VELLET, SESE EIVS NOMEN RECEPTVRVM, et simul ut ad causam accederet nomenque deferret, Agathinum coepit hortari. Tum ille clare omnibus audien-
60 tibus se id non esse facturum, neque se adeo Sthenio esse inimicum ut eum rei capitalis adfinem esse diceret. Hic tum repente Pacilius quidam, homo egens et levis, accedit; ait, si liceret, absentis nomen deferre se velle. Iste vero et licere et fieri solere,
65 et se recepturum; itaque defertur; edicit statim ut Kalendis Decembribus adsit Sthenius Syracusis.

§ 9

Meanwhile Sthenius laid his complaint before the Senate and Verres' father sent messengers to warn him; but Verres undeterred condemned Sthenius in the absence of both the accuser and the accused.

Hic qui Romam pervenisset, satisque feliciter anni iam adverso tempore navigasset, omniaque habuisset aequiora et placabiliora quam animum praetoris atque hospitis, rem ad amicos suos detulit,
5 quae, ut erat acerba atque indigna, sic videbatur omnibus. Itaque in senatu continuo Cn. Lentulus et L. Gellius consules faciunt mentionem placere statui, si patribus conscriptis videretur, ne absentes homines in provinciis rei fierent rerum capitalium;
10 causam Sthenii totam et istius crudelitatem et iniquitatem senatum docent. Aderat in senatu

Verres pater istius, et flens unum quemque senatorum rogabat ut filio suo parceret; neque tamen multum proficiebat; erat enim summa voluntas senatus. Itaque sententiae dicebantur: CVM STHE- 15 NIVS ABSENS REVS FACTVS ESSET, DE ABSENTE IVDICIVM NVLLVM FIERI PLACERE, ET, SI QVOD ESSET FACTVM, ID RATVM ESSE NON PLACERE. Eo die transigi nihil potuit, quod et id temporis erat et ille pater istius invenerat homines qui dicendo 20 tempus consumerent. Postea senex Verres defensores atque hospites omnes Sthenii convenit, rogat eos atque orat ne oppugnent filium suum, de Sthenio ne laborent; confirmat iis curaturum se esse ne quid ei per filium suum noceretur; se 25 homines certos eius rei causa in Siciliam et terra et mari esse missurum. Commoventur amici Sthenii, sperant fore ut patris litteris nuntiisque filius ab incepto revocetur. In senatu postea causa non agitur. Veniunt ad istum domestici nuntii, litteras 30 a patre afferunt ante Kalendas Decembres, cum isti etiam tum de Sthenio in integro tota res esset.

Hic iste, qui prae cupiditate neque officii sui neque periculi neque pietatis neque humanitatis rationem habuisset umquam, neque in eo, quod 35 monebatur, auctoritatem patris neque in eo, quod rogabatur, voluntatem anteponendam putavit libidini suae, mane Kalendis Decembribus, ut edixerat, Sthenium citari iubet. Citat reum: non respondet. Citat accusatorem. Citatus accusator, M. Pacilius, 40 nescio quo casu non respondit. Si praesens Sthenius reus esset factus, si manifesto in maleficio teneretur,

tamen, cum accusator non adesset, Sthenium condemnari non oporteret. Etenim, si posset reus
45 absente accusatore condemnari, non ego a Vibone Veliam parvulo navigio inter fugitivorum ac praedonum ac tua tela venissem, quo tempore omnis illa mea festinatio fuit cum periculo capitis ob eam causam, ne tu ex reis eximerere, si ego ad diem non
50 adfuissem. Ita fecit ut exitus principio simillimus reperiretur: quem absentem reum fecerat, eum absente accusatore condemnat.

§ 10

To assist Sthenius Verres appointed neither a friend nor a fellow-citizen but a Roman who was his open enemy. There he is—that sly-looking fellow over there, Verres' right-hand man!

Videte porro aliam amentiam; videte ut, dum expedire sese vult, induat. Cognitorem adscribit Sthenio—quem? cognatum aliquem aut propinquum? Non. Thermitanum aliquem, honestum
5 hominem ac nobilem? Ne id quidem. At Siculum, in quo aliquis splendor dignitasque esset? Neminem. Quid igitur? Civem Romanum. Cui hoc probari potest? Cum esset Sthenius civitatis suae nobilissimus, amplissima cognatione, plurimis ami-
10 citiis, cum praeterea tota Sicilia multum auctoritate et gratia posset, invenire neminem Siculum potuit qui pro se cognitor fieret? Hoc probabis?

At, credo, Sthenius hoc sibi amplum putavit, eligere e civium Romanorum numero, ex amicorum
15 atque hospitum suorum copia, quem cognitorem

daret. Quem delegit? Quis in tabulis scriptus est? C. Claudius Palatina. Non quaero quis hic sit Claudius, quam splendidus, quam honestus. Nihil horum quaero: fortasse enim Sthenius non splendorem hominis, sed familiaritatem secutus est. 20 Quid? si omnium mortalium nemo Sthenio inimicior quam hic C. Claudius cum semper tum in his ipsis rebus et temporibus fuit, si de litteris corruptis contra venit, si contra omni ratione pugnavit, utrum potius pro Sthenio inimicum cognitorem 25 esse factum an te ad Sthenii periculum inimici eius nomine abusum esse credemus? Ac ne quis forte dubitet cuius modi hoc totum sit negotium, tametsi iamdudum omnibus istius improbitatem perspicuam esse confido, tamen paulum etiam attendite. 30 Videtis illum subcrispo capillo, nigrum, qui eo vultu nos intuetur ut sibi ipsi peracutus esse videatur, qui tabulas tenet, qui scribit, qui monet, qui proximus est? Is est Claudius, qui in Sicilia sequester istius, interpres, confector negotiorum, 35 prope collega Timarchidi numerabatur, nunc obtinet eum locum ut vix Apronio illi de familiaritate concedere videatur, ei qui se non Timarchidi, sed ipsius Verris collegam et socium esse dicebat. Dubitate etiam, si potestis, quin eum iste potissi- 40 mum ex omni numero delegerit, cui hanc cognitoris falsi improbam personam imponeret, quem et huic inimicissimum et sibi amicissimum esse arbitraretur!

§ 11

Frightened by further news from Rome, Verres now erased the records of the case and so left himself defenceless. His previous claim that the trial had been legal was disproved when he falsified the records.

Nuntiabatur illi primis illis temporibus, id quod pater quoque ad eum pluribus verbis scripserat, agitatam rem esse in senatu. Cum haec ad istum adferrentur, pertimuit aliquando et commotus est;
5 vertit stilum in tabulis suis, quo facto causam omnem evertit suam; nihil enim sibi reliqui fecit quod defendi aliqua ratione posset. Nam si ita defenderet, "Recipi nomen absentis licet; hoc fieri in provincia nulla lex vetat", mala et improba
10 defensione, verum aliqua tamen uti videretur; postremo illo desperatissimo perfugio uti posset, se imprudentem fecisse, existimasse id licere. Quamquam haec perditissima defensio est, tamen aliquid dici videretur. Tollit ex tabulis id quod erat,
15 et facit coram esse delatum.

Hic videte in quot se laqueos induerit, quorum ex nullo se, iudices, umquam expediet. Primum ipse in Sicilia saepe et palam de loco superiore dixerat et in sermone multis demonstrarat licere nomen
20 recipere absentis; se exemplo fecisse quod fecisset. Deinde Romae, cum haec acta res esset in senatu, omnes istius amici, in his etiam pater eius hoc defendebat, licere fieri; saepe esse factum; iste quod fecisset aliorum exemplo institutoque fecisse.

25 Quae cum ita essent, tantane amentia praeditus atque audacia fuisti ut in re tam clara, tam testata,

tam abs te ipso pervulgata tabulas publicas corrumpere auderes? At quem ad modum corrupisti? nonne ita ut omnibus nobis tacentibus ipsae tuae te tabulae condemnare possent? Cedo, quaeso, 30 codicem, circumfer, ostende. Videtisne totum hoc nomen, coram ubi facit delatum, esse in litura? Quid fuit istic antea scriptum? quod mendum ista litura correxit? Quid a nobis, iudices, exspectatis argumenta huius criminis? Nihil dicimus; tabulae 35 sunt in medio, quae se corruptas atque interlitas esse clamant. Ex istis etiam tu rebus effugere te posse confidis, cum te nos non opinione dubia, sed tuis vestigiis persequamur, quae tu in tabulis publicis expressa ac recentia reliquisti? Is mihi 40 etiam Sthenium litteras publicas corrupisse causa incognita iudicavit, qui defendere non poterit se non in ipsius Sthenii nomine litteras publicas corrupisse?

PART II

VERRES THE COLLECTOR

§ 12

In the whole of Sicily there was no single work of art, whether in private or public hands, which Verres did not take if he fancied it.

Venio nunc ad istius, quem ad modum ipse appellat, studium, ut amici eius, morbum et insaniam, ut Siculi, latrocinium: ego quo nomine appellem nescio: rem vobis proponam; vos eam suo, non
5 nominis pondere penditote. Genus ipsum prius cognoscite, iudices; deinde fortasse non magno opere quaeretis quo id nomine appellandum putetis. Nego in Sicilia tota, tam locupleti, tam vetere provincia, tot oppidis, tot familiis tam copiosis,
10 ullum argenteum vas, ullum Corinthium aut Deliacum fuisse, ullam gemmam aut margaritam, quicquam ex auro aut ebore factum, signum ullum aeneum, marmoreum, eburneum, nego ullam picturam neque in tabula neque in textili quin con-
15 quisierit, inspexerit, quod placitum sit abstulerit. Cum dico nihil istum eius modi rerum in tota provincia reliquisse, Latine me scitote, non accusatorie loqui. Etiam planius: nihil in aedibus cuiusquam, ne in hospitis quidem; nihil in locis com-
20 munibus, ne in fanis quidem; nihil apud Siculum, nihil apud civem Romanum, denique nihil istum, quod ad oculos animumque acciderit, neque privati neque publici neque profani neque sacri tota in Sicilia reliquisse.

The Story of Heius

§ 13

Heius was the proud possessor of several statues by the most famous Greek sculptors. Verres took them all. The sanctity of the chapel in which they were kept could save them from any one else, but not from him.

C. Heius est Mamertinus omnibus rebus illa in civitate ornatissimus. Huius domus est vel optima Messanae, notissima quidem certe et nostris hominibus apertissima maximeque hospitalis. Ea domus ante istius adventum ornata sic fuit ut urbi esset ornamento. Erat apud Heium sacrarium magna cum dignitate in aedibus a maioribus traditum perantiquum, in quo signa pulcherrima quattuor summo artificio, summa nobilitate, quae non modo istum hominem ingeniosum et intellegentem, verum etiam quemvis nostrum, quos iste idiotas appellat, delectare possent, unum Cupidinis marmoreum Praxiteli; nimirum didici etiam, dum in istum inquiro, artificum nomina. Idem, opinor, artifex eiusdem modi Cupidinem fecit illum qui est Thespiis, propter quem Thespiae visuntur; nam alia visendi causa nulla est.

Verum ut ad illud sacrarium redeam, signum erat hoc quod dico Cupidinis e marmore, ex altera parte Hercules egregie factus ex aere. Is dicebatur esse Myronis, ut opinor, et certe. Item ante hos deos erant arulae, quae cuivis religionem sacrarii significare possent. Erant aenea duo praeterea signa, non maxima verum eximia venustate, virginali habitu atque vestitu, quae manibus sublatis

sacra quaedam more Atheniensium virginum reposita in capitibus sustinebant; Canephoroe ipsae vocabantur; sed earum artificem—quem? quemnam? recte admones—Polyclitum esse dicebant.
30 Messanam ut quisque nostrum venerat, haec visere solebat; omnibus haec ad visendum patebant cotidie; domus erat non domino magis ornamento quam civitati. Haec omnia quae dixi signa, iudices, ab Heio e sacrario Verres abstulit; nullum, inquam,
35 horum reliquit neque aliud ullum tamen praeter unum pervetus ligneum, Bonam Fortunam, ut opinor; eam iste habere domi suae noluit.

Pro deum hominumque fidem! quid hoc est? quae haec causa est, quae ista impudentia? Quae
40 dico signa, antequam abs te sublata sunt, Messanam cum imperio nemo venit quin viserit. Tot praetores, tot consules in Sicilia cum in pace tum etiam in bello fuerunt, tot homines cuiusque modi—non loquor de integris, innocentibus, religiosis—tot
45 cupidi, tot improbi, tot audaces, quorum nemo sibi tam vehemens, tam potens, tam nobilis visus est qui ex illo sacrario quicquam poscere aut tollere aut attingere auderet: Verres quod ubique erit pulcherrimum auferet? nihil habere cuiquam praeterea
50 licebit? tot domus locupletissimas istius domus una capiet? Idcirco nemo superiorum attigit ut hic tolleret? At non requirebat ille Cupido praedonis domum ac disciplinam; facile illo sacrario patrio continebatur; Heio se a maioribus relictum esse
55 sciebat in hereditate sacrorum, non quaerebat flagitiosum heredem.

§ 14

But Verres has a defence: he bought them! Yes, and for a song, according to his account books. Heius was not in need of money. His treasures were extorted from him.

Sed quid ego tam vehementer invehor? verbo uno repellar. "Emi", inquit. Di immortales, praeclaram defensionem! Mercatorem in provinciam cum imperio ac securibus misimus, omnia qui signa, tabulas pictas, omne argentum, aurum, ebur, 5 gemmas coemeret, nihil cuiquam relinqueret!

Videamus quanta ista pecunia fuerit quae potuerit Heium, hominem maxime locupletem, minime avarum, ab humanitate, a pietate, ab religione deducere. Ita iussisti, opinor, ipsum in tabulas 10 referre: "Haec omnia signa Praxiteli, Myronis, Polycliti HS sex milibus quingentis Verri vendita". Sic rettulit. Iuvat me haec praeclara nomina artificum, quae isti ad caelum ferunt, Verris aestimatione sic concidisse. Cupidinem Praxiteli HS 15 MDC! Profecto hinc natum est, "Malo emere quam rogare". Video igitur Heium neque voluntate neque difficultate aliqua temporis nec magnitudine pecuniae adductum esse ut haec signa venderet, teque ista simulatione emptionis vi, metu, imperio, 20 fascibus ab homine eo quem, una cum ceteris sociis, non solum potestati tuae sed etiam fidei populus Romanus commiserat eripuisse atque abstulisse.

§ 15

Verres had brought with him two rascally agents, fugitives from justice, whom he used for discovering and acquiring works of art.

Iam, ut haec omnia reperire ac perscrutari solitus sit, iudices, est operae pretium cognoscere. Cibyratae sunt fratres quidam, Tlepolemus et Hiero, quorum alterum fingere opinor e cera solitum esse, alterum
5 esse pictorem. Hosce opinor, Cibyrae cum in suspicionem venissent suis civibus fanum expilasse Apollinis, veritos poenam iudicii ac legis domo profugisse. Quod Verrem artificii sui cupidum cognoverant, domo fugientes ad eum se exsules
10 contulerunt. Eos iam bene cognitos et re probatos secum in Siciliam duxit. Quo posteaquam venerunt, mirandum in modum (canes venaticos diceres) ita odorabantur omnia et pervestigabant ut, ubi quidque esset, aliqua ratione invenirent. Aliud minando,
15 aliud pollicendo, aliud per servos, aliud per liberos, per amicum aliud, aliud per inimicum inveniebant; quicquid illis placuerat, perdendum erat. Nihil aliud optabant quorum poscebatur argentum nisi ut id Hieroni et Tlepolemo displiceret.

§ 16

They approached Pamphilus, who had already suffered one robbery and was threatened with another, and asked him how much he would pay for their good offices. Pamphilus had to give them a large bribe to save his goblets; for Verres acted on their judgment.

Verum mehercule hoc, iudices, dicam. Memini Pamphilum Lilybitanum, amicum et hospitem meum, mihi narrare, cum iste ab sese hydriam Boethi manu factam praeclaro opere et grandi pondere per potestatem abstulisset, se sane tristem et contur- 5 batum domum revertisse, quod vas eius modi, quod sibi a patre et a maioribus esset relictum, quo solitus esset uti ad festos dies, ad hospitum adventus, a se esset ablatum. "Cum sederem", inquit, "domi tristis, accurrit Venerius; iubet me scyphos sigil- 10 latos ad praetorem statim adferre. Permotus sum," inquit; "binos habebam; iubeo promi utrosque, ne quid plus mali nasceretur, et mecum ad praetoris domum ferri. Eo cum venio, praetor quiescebat; fratres illi Cibyratae inambulabant. Qui me ubi 15 viderunt, 'Ubi sunt, Pamphile,' inquiunt, 'scyphi?' Ostendo tristis; laudant. Incipio queri me nihil habiturum quod alicuius esset pretii si etiam scyphi essent ablati. Tum illi, ubi me conturbatum vident, 'Quid vis nobis dare ut isti abs te ne auferantur?' 20 Ne multa, HS mille me", inquit, "poposcerunt; dixi me daturum. Vocat interea praetor, poscit scyphos." Tum illos coepisse praetori dicere putasse se, id quod audissent, alicuius pretii scyphos esse Pamphili; luteum negotium esse, non dignum 25 quod in suo argento Verres haberet. Ait ille idem sibi videri. Ita Pamphilus scyphos optimos aufert.

Et mehercule ego antea, tametsi hoc nescio quid nugatorium sciebam esse, ista intellegere, tamen 30 mirari solebam istum in his ipsis rebus aliquem sensum habere, quem scirem nulla in re quicquam simile hominis habere. Tum primum intellexi ad eam rem istos fratres Cibyratas fuisse, ut iste in furando manibus suis, oculis illorum uteretur.

THE STORY OF PRINCE ANTIOCHUS

§ 17

Passing over lesser instances, I come to his greatest outrage. A Syrian prince, visiting Sicily on his way back from Rome, fell into the hands of Verres, who coveted his plate, and borrowed some of it.

Iam illa quae leviora videbuntur ideo praeteribo, quod mensas Delphicas e marmore, crateras ex aere pulcherrimas, vim maximam vasorum Corinthiorum ex omnibus aedibus sacris abstulit Syracusis. 5 Itaque, iudices, ii qui hospites ad ea quae visenda sunt solent ducere et unum quidque ostendere,—quos illi mystagogos vocant,—conversam iam habent demonstrationem suam. Nam ut ante demonstrabant quid ubique esset, item nunc quid 10 undique ablatum sit ostendunt.

Venio nunc non iam ad furtum, non ad avaritiam, non ad cupiditatem, sed ad eius modi facinus in quo omnia nefaria contineri mihi atque inesse videantur; in quo di immortales violati, existimatio atque auctoritas nominis populi Romani imminuta, hospitium spoliatum ac proditum, abalienati scelere istius a nobis omnes reges amicissimi, nationesque quae in eorum regno ac dicione sunt. Nam reges Syriae, regis Antiochi filios pueros, scitis Romae nuper fuisse; qui posteaquam temporibus rei publicae exclusi per senatum agere quae voluerant non potuerunt, in Syriam in regnum patrium profecti sunt. Eorum alter, qui Antiochus vocatur, iter per Siciliam facere voluit, itaque isto praetore venit Syracusas. 15 20 25

Hic Verres hereditatem sibi venisse arbitratus est, quod in eius regnum ac manus venerat is quem iste et audierat multa secum praeclara habere et suspicabatur. Mittit homini munera satis large ad usum domesticum. Deinde ipsum regem ad cenam vocavit. Exornat ample magnificeque triclinium; exponit ea, quibus abundabat, plurima et pulcherrima vasa argentea,—nam haec aurea nondum fecerat; omnibus curat rebus instructum et paratum ut sit convivium. Quid multa? rex ita discessit ut et istum copiose ornatum et se honorifice acceptum arbitraretur. Vocat ad cenam deinde ipse praetorem; exponit suas copias omnes, multum argentum, non pauca etiam pocula ex auro, quae, ut mos est regius et maxime in Syria, gemmis erant distincta clarissimis. Erat etiam vas vinarium, ex una gemma pergrandi trulla excavata, manubrio aureo. Iste unum quodque vas in manus sumere, laudare, mirari: rex gaudere praetori populi Romani satis iucundum et gratum illud esse convivium. Postea- 30 35 40 45

quam inde discessum est, cogitare nihil iste aliud, quod ipsa res declaravit, nisi quem ad modum regem ex provincia spoliatum expilatumque dimitteret. Mittit rogatum vasa ea quae pulcherrima
50 apud eum viderat; ait se suis caelatoribus velle ostendere. Rex, qui illum non nosset, sine ulla suspicione libentissime dedit. Mittit etiam trullam gemmeam rogatum; velle se eam diligentius considerare. Ea quoque ei mittitur.

§ 18

In particular, he borrowed a superb candelabrum which Antiochus meant to present to the new temple of Jupiter Optimus Maximus at Rome, and refused to give it back. The prince, despite his public protests, he ejected from Sicily.

Nunc reliquum, iudices, attendite. Candelabrum e gemmis clarissimis opere mirabili perfectum reges ii, quos dico, Romam cum attulissent, ut in Capitolio ponerent, quod nondum perfectum tem-
5 plum offenderant, neque ponere potuerunt neque vulgo ostendere ac proferre voluerunt; statuerunt id secum in Syriam reportare ut, cum audissent simulacrum Iovis Optimi Maximi dedicatum, legatos mitterent qui cum ceteris rebus illud quoque
10 eximium ac pulcherrimum donum in Capitolium adferrent. Pervenit res ad istius aures nescio quo modo. Iste petit a rege et eum pluribus verbis rogat ut id ad se mittat; cupere se dicit inspicere neque se aliis videndi potestatem esse facturum. Antio-
15 chus, qui animo et puerili esset et regio, nihil de

istius improbitate suspicatus est; imperat suis ut id in praetorium involutum quam occultissime deferrent. Quo posteaquam attulerunt involucrisque reiectis constituerunt, clamare iste coepit dignam rem esse regno Syriae, dignam regio 20 munere, dignam Capitolio. Cum satis iam perspexisse videretur, tollere incipiunt ut referrent. Iste ait se velle illud etiam atque etiam considerare; nequaquam se esse satiatum; iubet illos discedere et candelabrum relinquere. Sic illi tum inanes ad 25 Antiochum revertuntur.

Rex primo nihil metuere, nihil suspicari; dies unus, alter, plures; non referri. Tum mittit, si videatur, ut reddat. Iubet iste posterius ad se reverti. Mirum illi videri; mittit iterum; non 30 redditur. Ipse hominem appellat, rogat ut reddat. Os hominis insignemque impudentiam cognoscite. Quod sciret, quod ex ipso rege audisset in Capitolio esse ponendum, quod Iovi Optimo Maximo, quod populo Romano servari videret, id sibi ut donaret 35 rogare et vehementissime petere coepit. Cum ille se et religione Iovis Capitolini et hominum existimatione impediri diceret, quod multae nationes testes essent illius operis ac muneris, iste homini minari acerrime coepit. Ubi videt eum nihilo magis 40 minis quam precibus permoveri, repente hominem de provincia iubet ante noctem decedere; ait se comperisse ex eius regno piratas ad Siciliam esse venturos. Rex maximo conventu Syracusis in foro flens ac deos hominesque contestans clamare coepit 45 candelabrum factum e gemmis, quod in Capitolium

missurus esset, quod in templo clarissimo populo Romano monumentum suae societatis amicitiaeque esse voluisset, id sibi C. Verrem abstulisse; de ceteris
50 operibus ex auro et gemmis quae sua penes illum essent se non laborare, hoc sibi eripi miserum esse et indignum. Id etsi antea iam mente et cogitatione sua fratrisque sui consecratum esset, tamen tum se in illo conventu civium Romanorum dare donare
55 dicare consecrare Iovi Optimo Maximo, testemque ipsum Iovem suae voluntatis ac religionis adhibere.

§ 19

What an outrage to a friendly prince, to the feelings of foreigners, to the honour of Rome!

Quae vox, quae latera, quae vires huius unius criminis querimoniam possunt sustinere? Rex Antiochus, cum amicus et socius populi Romani esset, amicissimo patre, avo, maioribus, antiquis-
5 simis et clarissimis regibus, opulentissimo et maximo regno, praeceps provincia populi Romani exturbatus est. Quem ad modum hoc accepturas nationes exteras putasti, cum audirent a praetore populi Romani in provincia violatum regem, spoliatum
10 hospitem, eiectum socium populi Romani atque amicum? Nomen vestrum populique Romani odio atque acerbitati scitote nationibus exteris, iudices, futurum, si istius haec tanta iniuria impunita discesserit.

15 Vobis autem, iudices, quid hoc indignius aut quid minus ferendum videri potest? Verresne

habebit domi suae candelabrum Iovis e gemmis auroque perfectum? cuius fulgore conlucere atque inlustrari Iovis Optimi Maximi templum oportebat, id apud istum in conviviis constituetur? 20

The Story of Sopater

§ 20

After this he stopped at nothing. At Tyndaris he ordered Sopater to remove the statue of Mercury. It was impossible for him to obey despite Verres' threats.

Itaque hoc nefario scelere concepto nihil postea tota in Sicilia neque sacri neque religiosi duxit esse; ita sese in ea provincia per triennium gessit ut ab isto non solum hominibus verum etiam dis immortalibus bellum indictum putaretur. 5

A Tyndaritanis non simulacrum Mercurii pulcherrime factum sustulisti? At quem ad modum, di immortales! quam audacter, quam libidinose, quam impudenter! Audistis nuper dicere legatos Tyndaritanos Mercurium, qui sacris anniversariis 10 ac summa religione coleretur, huius vi scelere imperioque esse sublatum. Qui ut primum in illud oppidum venit, statim, tamquam ita fieri non solum oporteret sed etiam necesse esset, tamquam hoc senatus mandasset populusque Romanus iussisset, 15 ita continuo signum ut demolirentur et Messanam deportarent imperavit. Quod cum illis qui aderant indignum, qui audiebant incredibile videretur, non est ab isto primo illo adventu perseveratum.

20 Discedens mandat proagoro Sopatro, cuius verba audistis, ut demoliatur; cum recusaret, vehementer minatur et statim ex illo oppido proficiscitur. Refert rem ille ad senatum; vehementer undique reclamatur. Ne multa, iterum iste ad illos aliquanto
25 post venit, quaerit continuo de signo. Respondetur ei senatum non permittere; poenam capitis constitutam, si iniussu senatus quisquam attigisset; simul religio commemoratur. Tum iste, "Quam mihi religionem narras, quam poenam, quem
30 senatum? vivum te non relinquam; moriere virgis nisi mihi signum traditur". Sopater iterum flens ad senatum rem defert, istius cupiditatem minasque demonstrat. Senatus Sopatro responsum nullum dat, sed commotus perturbatusque discedit. Ille
35 praetoris arcessitus nuntio rem demonstrat, negat ullo modo fieri posse. Atque haec—nihil enim praetermittendum de istius impudentia videtur—agebantur in conventu palam de sella ac de loco superiore.

§ 21

Thereupon Verres had him stripped and bound, naked, on a bronze equestrian statue in the forum. The townspeople, fearing that he would die of exposure, gave way.

Erat hiems summa, tempestas, ut ipsum Sopatrum dicere audistis, perfrigida, imber maximus, cum iste imperat lictoribus ut Sopatrum de porticu, in qua ipse sedebat, praecipitem in forum deiciant
5 nudumque constituant. Vix erat hoc plane imperatum cum illum spoliatum stipatumque lictoribus

videres. Omnes id fore putabant ut miser atque innocens virgis caederetur; fefellit hic homines opinio. Leniter hominem clementerque accepit. Equestres sunt medio in foro Marcellorum statuae, 10 sicut fere ceteris in oppidis Siciliae; ex quibus iste C. Marcelli statuam delegit, cuius officia in illam civitatem totamque provinciam recentissima erant et maxima; in ea Sopatrum, hominem cum domi nobilem tum summo magistratu praeditum, divari- 15 cari ac deligari iubet. Quo cruciatu sit affectus venire in mentem necesse est omnibus, cum esset vinctus nudus in aere, in imbri, in frigore. Neque tamen finis huic iniuriae crudelitatique fiebat donec populus atque universa multitudo, atrocitate rei 20 misericordiaque commota, senatum clamore coegit ut isti simulacrum illud Mercurii polliceretur. Clamabant fore ut ipsi se di immortales ulciscerentur; hominem interea perire innocentem non oportere. Tum frequens senatus ad istum venit, 25 pollicetur signum. Ita Sopater de statua C. Marcelli, cum iam paene obriguisset, vix vivus aufertur.

The Spoliation of Syracuse

§ 22

Syracuse, spared in war by its captor, was pillaged in peace by its governor.

Unius etiam urbis omnium pulcherrimae atque ornatissimae, Syracusarum, direptionem commemorabo et in medium proferam, iudices, ut aliquando totam huius generis orationem concludam

5 atque definiam. Nemo fere vestrum est quin quem ad modum captae sint a M. Marcello Syracusae saepe audierit, non numquam etiam in annalibus legerit. Conferte hanc pacem cum illo bello, huius praetoris adventum cum illius imperatoris victoria,
10 huius cohortem impuram cum illius exercitu invicto, huius libidines cum illius continentia: ab illo qui cepit conditas, ab hoc qui constitutas accepit captas dicetis Syracusas. Ac iam illa omitto quae disperse a me multis in locis dicentur ac dicta sunt,
15 forum Syracusanorum, quod introitu Marcelli purum a caede servatum est, id adventu Verris Siculorum innocentium sanguine redundasse, portum Syracusanorum, qui tum et nostris classibus et Carthaginiensium clausus fuisset, eum isto praetore
20 Cilicum myoparoni praedonibusque patuisse; mitto, inquam, haec omnia, quae ab isto per triennium perfecta sunt; ea quae coniuncta cum illis rebus sunt de quibus antea dixi cognoscite.

§ 23

He robbed the temple of Minerva of its famous pictures and stripped its doors, without a thought that he might be called to account.

Aedis Minervae est in Insula; quam Marcellus non attigit, quam plenam atque ornatam reliquit; quae ab isto sic spoliata atque direpta est non ut ab hoste aliquo, qui tamen in bello religionem et consuetu-
5 dinis iura retineret, sed ut a barbaris praedonibus vexata esse videatur. Pugna erat equestris Agathocli regis in tabulis picta praeclare; iis autem

tabulis interiores templi parietes vestiebantur. Nihil erat ea pictura nobilius, nihil Syracusis quod magis visendum putaretur. Has tabulas M. Marcellus, cum omnia victoria illa sua profana fecisset, tamen religione impeditus non attigit; iste, cum illa propter diuturnam pacem fidelitatemque populi Syracusani sacra religiosaque accepisset, omnes eas tabulas abstulit, parietes, quorum ornatus tot saecula manserant, tot bella effugerant, nudos ac deformatos reliquit.

Iam vero quid ego de valvis illius templi commemorem? Confirmare hoc liquido, iudices, possum, valvas magnificentiores, ex auro atque ebore perfectiores, nullas umquam ullo in templo fuisse. Ex ebore diligentissime perfecta argumenta erant in valvis; ea detrahenda curavit omnia. Gorgonis os pulcherrimum cinctum anguibus revellit atque abstulit, et tamen indicavit se non solum artificio sed etiam pretio quaestuque duci; nam bullas aureas omnes ex iis valvis, quae erant multae et graves, non dubitavit auferre; quarum iste non opere delectabatur sed pondere. Itaque eius modi valvas reliquit ut quae olim ad ornandum templum erant maxime, nunc tantum ad claudendum factae esse videantur.

Neque ego nunc istius facta omnia enumerare conor, neque opus est nec fieri ullo modo potest: tantum unius cuiusque de varia improbitate generis indicia apud vos et exempla profero. Neque enim ita se gessit in his rebus tamquam rationem aliquando esset redditurus, sed prorsus ita quasi

aut reus numquam esset futurus, aut, quo plura
40 abstulisset, eo minore periculo in iudicium venturus
esset; qui haec quae dico iam non occulte, non per
amicos atque interpretes, sed palam de loco superiore ageret pro imperio et potestate.

PART III

VERRES THE GENERAL

§ 24

But Verres relies on his reputation as a general to cover his misdeeds as a governor. Well, what sort of a general was this "saviour of the country"?

Nemini video dubium esse, iudices, quin apertissime
C. Verres in Sicilia sacra profanaque omnia et
privatim et publice spoliarit, versatusque sit sine
ulla non modo religione verum etiam dissimulatione
in omni genere furandi atque praedandi. Sed 5
quaedam mihi magnifica et praeclara eius defensio
ostenditur; cui quem ad modum resistam multo
mihi ante est, iudices, providendum. Ita enim
causa constituitur, provinciam Siciliam virtute
istius et vigilantia singulari dubiis formidolosisque 10
temporibus a belli periculis tutam esse servatam.
Quid agam, iudices? quo accusationis meae rationem conferam? quo me vertam? ad omnes enim
meos impetus quasi murus quidam boni nomen
imperatoris opponitur. Novi locum; video ubi se 15
iactaturus sit Hortensius. Belli pericula, tempora
rei publicae, imperatorum penuriam commemo-
rabit; tum deprecabitur a vobis, tum etiam pro suo
iure contendet ne patiamini talem imperatorem
populo Romano Siculorum testimoniis eripi, ne 20
obteri laudem imperatoriam criminibus avaritiae
velitis.

Non possum dissimulare, iudices; timeo ne C. Verres propter hanc eximiam virtutem in re
25 militari omnia quae fecit impune fecerit. Sit fur, sit sacrilegus, sit flagitiorum omnium vitiorumque princeps, at est bonus imperator, at felix et ad dubia rei publicae tempora reservandus. Cupio mihi ab illo, iudices, subici, quoniam de militari
30 eius gloria dico, si quid forte praetereo; certe nihil sciens praetermittam. Habetis hominis consilia, diligentiam, vigilantiam, custodiam defensionemque provinciae. Summa illuc pertinet, ut sciatis, quoniam plura genera sunt imperatorum, ex quo
35 genere iste sit, ne qui diutius in tanta penuria virorum fortium talem imperatorem ignorare possit.

§ 25

The inconveniences of winter travel he got over by staying in his luxurious quarters at Syracuse. With the spring, he issued out in his litter and dispensed venal justice from his bedroom. His riotous banquets were a public scandal.

Itinerum primum laborem, qui vel maximus est in re militari, iudices, et in Sicilia maxime necessarius, accipite quam facilem sibi iste et iucundum ratione consilioque reddiderit. Primum temporibus
5 hibernis ad magnitudinem frigorum et tempestatum vim ac fluminum praeclarum hoc sibi remedium compararat. Urbem Syracusas elegerat, cuius hic situs atque haec natura esse loci caelique dicitur ut nullus umquam dies tam magna ac turbulenta
10 tempestate fuerit quin aliquo tempore eius diei

solem homines viderint. Hic ita vivebat iste bonus imperator hibernis mensibus ut eum non facile non modo extra tectum, sed ne extra lectum quidem quisquam viderit; ita diei brevitas conviviis, noctis longitudo flagitiis continebatur. 15

Cum autem ver esse coeperat—cuius initium iste non a Favonio neque ab aliquo astro notabat, sed cum rosam viderat tum incipere ver arbitrabatur—dabat se labori atque itineribus; in quibus eo usque se praebebat patientem atque impigrum ut eum 20 nemo umquam in equo sedentem viderit. Nam, ut mos fuit Bithyniae regibus, lectica octaphoro ferebatur, in qua pulvinus erat perlucidus Melitensis rosa fartus; ipse autem coronam habebat unam in capite, alteram in collo, reticulumque ad nares sibi 25 admovebat tenuissimo lino, minutis maculis, plenum rosae. Sic confecto itinere cum ad aliquod oppidum venerat, eadem lectica usque in cubiculum deferebatur. Eo veniebant Siculorum magistratus, veniebant equites Romani, id quod ex multis 30 iuratis audistis; controversiae secreto deferebantur, paulo post palam decreta auferebantur. Deinde ubi paulisper in cubiculo pretio non aequitate iura discripserat, Libero iam reliquum tempus deberi arbitrabatur. 35

Erant vero convivia non illo silentio populi Romani praetorum atque imperatorum, neque eo pudore qui in magistratuum conviviis versari soleat, sed cum maximo clamore atque convicio; non numquam etiam res ad pugnam atque ad manus 40 vocabatur. Iste enim praetor severus ac diligens,

qui populi Romani legibus numquam paruisset, illis legibus quae in poculis ponebantur diligenter obtemperabat. Itaque erant exitus eius modi ut
45 alius inter manus e convivio tamquam e proelio auferretur, alius tamquam occisus relinqueretur, plerique ut fusi sine mente ac sine ullo sensu iacerent,—ut quivis, cum aspexisset, non se praetoris convivium, sed Cannensem pugnam nequitiae
50 videre arbitraretur.

Ac per eos dies, cum iste cum pallio purpureo talarique tunica in conviviis versaretur, non offendebantur homines neque moleste ferebant abesse a foro magistratum, non ius dici, non iudicia fieri;
55 locum illum litoris percrepare totum mulierum vocibus cantuque symphoniae, in foro silentium esse summum causarum atque iuris, non ferebant homines moleste; non enim ius abesse videbatur a foro neque iudicia, sed vis et crudelitas et bonorum
60 acerba et indigna direptio.

§ 26

So much for his claims to military renown! There was no war in Sicily. There was war outside it, against the pirates, but Verres turned pirate himself and embezzled his sailors' pay.

Hunc tu igitur imperatorem esse defendis, Hortensi? huius furta, rapinas, cupiditatem, crudelitatem, superbiam, scelus, audaciam rerum gestarum magnitudine atque imperatoriis laudibus tegere conaris?
5 Esto; nihil ex bello aut suspicione belli laudis

adeptus est, quod neque bellum neque belli periculum fuit in Sicilia, neque ab isto provisum est ne quod esset; at vero contra bellum praedonum classem habuit ornatam diligentiamque in eo singularem, itaque ab isto praeclare defensa provincia 10 est. Sic de bello praedonum, sic de classe Siciliensi, iudices, dicam ut hoc iam ante confirmem, in hoc uno genere omnes inesse culpas istius maximas avaritiae, maiestatis, dementiae, libidinis, crudelitatis. Haec dum breviter expono, quaeso, ut 15 fecistis adhuc, diligenter attendite.

Cognoscite enim novam praedandi rationem ab hoc primum excogitatam. Sumptum omnem in classem frumento stipendio ceterisque rebus suo quaeque nauarcho civitas semper dare solebat. 20 Verres post imperium constitutum primus imperavit ut ea pecunia omnis a civitatibus sibi adnumeraretur, ut is eam pecuniam tractaret quem ipse praefecisset. Deinde alii quaestus instituuntur, ex uno genere navali videte quam multi! accipere 25 a civitatibus pecuniam ne nautas darent, pretio certo missos facere nautas, missorum omne stipendium lucrari, reliquis quod deberet non dare,—haec omnia ex civitatum testimoniis cognoscite. Huncine hominem, hancine impudentiam, iudices, hanc 30 audaciam! civitatibus pro numero militum pecuniarum summas discribere, certum pretium, sescenos nummos, nautarum missionis constituere!

§ 27

A pirate ship, caught unawares by his under-manned fleet, he treated as his perquisite. He disposed of its crew for his own profit, executing Roman citizens in their place. This is the only victory he has to his credit.

Cum propter istius hanc avaritiam nomine classis
esset in Sicilia, re quidem vera naves inanes, quae
praedam praetori non quae praedonibus metum
adferrent, tamen, cum P. Caesetius et P. Tadius
5 decem navibus suis semiplenis navigarent, navem
quandam piratarum praeda refertam non ceperunt,
sed abduxerunt onere suo plane captam atque
depressam. Quod ubi isti nuntiatum est, tametsi
in acta iacebat ebrius, erexit se tamen et statim
10 quaestori legatoque suo custodes misit complures,
ut omnia sibi integra quam primum exhiberentur.
Adpellitur navis Syracusas; exspectatur ab omnibus
supplicium. Iste quasi praeda sibi advecta, non
praedonibus captis, si qui senes erant, eos in hos-
15 tium numero ducit; qui aliquid artificii habebant,
abducit omnes, non nullos scribis filio cohortique
distribuit, symphoniacos homines sex cuidam
amico suo Romam muneri misit. Nox illa tota in
exinaniunda nave consumitur. Archipiratam ipsum
20 videt nemo, de quo supplicium sumi oportuit.
Hodie omnes sic habent istum clam a piratis ob
hunc archipiratam pecuniam accepisse.

Interea Syracusani, homines periti et humani,
qui non modo ea quae perspicua essent videre
25 verum etiam occulta suspicari possent, habebant
rationem omnes cotidie piratarum qui securi feri-

rentur; quam multos esse oporteret, ex ipso navigio quod erat captum et ex remorum numero coniciebant. Cum magnus numerus deesset, tum iste homo nefarius in eorum locum quos domum suam 30 de piratis abduxerat substituere et supponere coepit cives Romanos, quos in carcerem antea coniecerat. Itaque alii cives Romani, ne cognoscerentur, capitibus obvolutis e carcere ad palum atque ad necem rapiebantur, alii, cum a multis civibus Romanis 35 cognoscerentur, ab omnibus defenderentur, securi feriebantur. Haec igitur est gesta res, haec victoria praeclara: myoparone piratico capto dux liberatus, symphoniaci Romam missi, artifices domum abducti, in eorum locum et ad eorum numerum cives 40 Romani hostilem in modum cruciati et necati, omnis vestis ablata, omne aurum et argentum ablatum et aversum.

§ 28

The summer Verres spent, not at sea, but under canvas, leaving Cleomenes to command the fleet.

Hac tanta praeda auctus, mancipiis argento veste locupletatus, nihilo diligentior ad classem ornandam milites revocandos alendosque esse coepit, cum ea res non solum provinciae saluti verum etiam ipsi praedae posset esse. Nam aestate summa, quo 5 tempore ceteri praetores obire provinciam et concursare consuerunt aut etiam in tanto praedonum metu et periculo ipsi navigare, eo tempore ad luxuriem suam domo sua regia contentus non fuit;

10 tabernacula carbaseis intenta velis conlocari iussit in litore, propter ipsum introitum atque ostium portus amoeno sane et ab arbitris remoto loco. Hic dies aestivos vivit praetor populi Romani, custos defensorque provinciae; naves quibus legatus prae-
15 fuerat Cleomeni tradit, classi populi Romani Cleomenem Syracusanum praeesse iubet atque imperare.

§ 29

The crews were on shore, foraging, and the admiral was drunk, when news of the pirate fleet was received. Cleomenes, in the fastest ship, spread his sails and fled. The rest followed, but were overtaken and captured or burnt.

Egreditur in Centuripina quadriremi Cleomenes e portu; sequitur Segestana navis, Tyndaritana, Herbitensis, Heracliensis, Apolloniensis, Haluntina, praeclara classis in speciem, sed inops et infirma
5 propter dimissionem propugnatorum atque remigum. Tam diu in imperio suo classem iste praetor diligens vidit quam diu convivium eius flagitiosissimum praetervecta est; ipse autem, qui visus multis diebus non esset, tum se tamen in conspec-
10 tum nautis paulisper dedit. Stetit soleatus praetor populi Romani cum pallio purpureo tunicaque talari in litore.

Posteaquam paulum provecta classis est et Pachynum quinto die denique adpulsa, nautae
15 coacti fame radices palmarum agrestium, quarum erat in illis locis, sicuti in magna parte Siciliae, multitudo, colligebant et iis miseri perditique ale-

bantur; Cleomenes autem, qui alterum se Verrem cum luxurie ac nequitia tum etiam imperio putaret, similiter totos dies in litore tabernaculo posito 20 perpotabat. Ecce autem repente ebrio Cleomene esurientibus ceteris nuntiatur piratarum esse naves in portu Odysseae; nam ita is locus nominatur; nostra autem classis erat in portu Pachyni.

Princeps Cleomenes in quadriremi Centuripina 25 malum erigi, vela fieri, praecidi ancoras imperavit, et simul ut se ceteri sequerentur signum dari iussit. Haec Centuripina navis erat incredibili celeritate velis. Evolarat iam e conspectu fere fugiens quadriremis, cum etiam tum ceterae naves uno in loco 30 moliebantur. Erat animus in reliquis; quamquam erant pauci, quoquo modo res se habebat, pugnare tamen se velle clamabant, et quod reliquum vitae viriumque fames fecerat id ferro potissimum reddere volebant. Quodsi Cleomenes non tanto ante 35 fugisset, aliqua tamen ad resistendum ratio fuisset. Erat enim sola illa navis constrata et ita magna ut propugnaculo ceteris posset esse; quae si in praedonum pugna versaretur, urbis instar habere inter illos piraticos myoparones videretur; sed tum 40 inopes, relicti ab duce praefectoque classis, eundem necessario cursum tenere coeperunt. Helorum versus, ut ipse Cleomenes, ita ceteri navigabant, neque ii tam praedonum impetum fugiebant quam imperatorem sequebantur. Tum ut quisque in fuga 45 postremus, ita in periculo princeps erat; postremam enim quamque navem piratae primam adoriebantur. Ita prima Haluntinorum navis capitur;

deinde Apolloniensis et eius praefectus Anthropinus
50 occiditur.

Haec dum aguntur, interea Cleomenes iam ad Helori litus pervenerat; iam sese in terram e navi eiecerat quadrirememque fluctuantem in salo reliquerat. Reliqui praefecti navium, cum in terram
55 imperator exisset, cum ipsi neque repugnare neque mari effugere ullo modo possent, adpulsis ad Helorum navibus Cleomenem persecuti sunt. Tum praedonum dux Heracleo, repente praeter spem non sua virtute sed istius avaritia nequitiaque
60 victor, classem pulcherrimam populi Romani in litus expulsam et eiectam, cum primum invesperasceret, inflammari incendique iussit.

§ 30

The news reached Syracuse at night, but no one dared wake Verres. He was roused at last by the execrations of the mob. The pirates entered the harbour unopposed.

O tempus miserum atque acerbum provinciae Siciliae! o casum illum multis innocentibus calamitosum atque funestum! o istius nequitiam ac turpitudinem singularem! Adfertur nocte intem-
5 pesta gravis huiusce mali nuntius Syracusas; curritur ad praetorium, quo istum e praeclaro convivio reduxerant paulo ante sodales cum cantu atque symphonia. Cleomenes, quamquam nox erat, tamen in publico esse non audet; includit se domi. Huius
10 autem praeclari imperatoris ita erat severa domi disciplina ut in re tanta et tam gravi nuntio nemo

admitteretur, nemo esset qui auderet aut dormientem excitare aut interpellare vigilantem.

Iam vero re ab omnibus cognita concursabat urbe tota maxima multitudo. Non enim, sicut erat 15 antea semper consuetudo, praedonum adventum significabat ignis e specula sublatus aut tumulo, sed flamma ex ipso incendio navium et calamitatem acceptam et periculum reliquum nuntiabat. Cum praetor quaereretur et constaret neminem ei nun- 20 tiasse, fit ad domum eius cum clamore concursus atque impetus. Tum iste excitatus audit rem omnem ex Timarchide, sagum sumit,—lucebat iam fere,—procedit in medium vini somni crapulae plenus. Excipitur ab omnibus strepitu ac clamore. 25 Tum istius acta commemorabantur, tum flagitiosa illa convivia, tum quaerebant ex isto palam tot dies continuos per quos numquam visus esset ubi fuisset, quid egisset.

Unam illam noctem solam praedones ad Helorum 30 commorati, cum fumantes etiam nostras naves reliquissent, accedere incipiunt Syracusas; qui videlicet saepe audissent nihil esse pulchrius quam Syracusarum moenia ac portus, statuerant se, si ea Verre praetore non vidissent, numquam esse visuros. 35 Ac primo ad illa aestiva praetoris accedunt, ipsam illam ad partem litoris ubi iste per eos dies tabernaculis positis castra luxuriae conlocarat. Quem posteaquam inanem locum offenderunt et praetorem commosse ex eo loco castra senserunt, statim 40 sine ullo metu in ipsum portum penetrare coeperunt. Hic te praetore Heracleo pirata cum quattuor

myoparonibus parvis ad arbitrium suum navigavit. Pro di immortales! piraticus myoparo, cum imperii
45 populi Romani nomen ac fasces essent Syracusis, usque ad forum Syracusanorum et ad omnes crepidines urbis accessit.

§ 31

For this the blame was openly laid on Verres, who had cheated the sailors of their pay and even their rations. What a disgrace to Sicily and Rome!

Posteaquam e portu piratae non metu aliquo adfecti sed satietate exierunt, tum coeperunt quaerere homines causam illius tantae calamitatis. Dicere omnes et palam disputare minime esse mirandum
5 si remigibus militibusque dimissis, reliquis egestate et fame perditis, praetore tot dies cum sodalibus perpotante, tanta ignominia et calamitas esset accepta. Haec autem istius vituperatio atque infamia confirmabatur eorum sermone qui a suis
10 civitatibus illis navibus praepositi fuerant. Qui ex illo numero reliqui Syracusas classe amissa refugerant dicebant quot ex sua quisque nave missos sciret esse. Res erat clara, neque solum argumentis sed etiam certis testibus istius audacia tenebatur.
15 Quid ais, bone custos defensorque provinciae? Eone pirata penetravit quo simul atque adisset non modo a latere sed etiam a tergo magnam partem urbis relinqueret? Insulam totam praetervectus est, quae est urbs Syracusis suo nomine ac moeni-
20 bus, quo in loco maiores Syracusanum habitare

vetuerunt, quod, qui illam partem urbis tenerent, in eorum potestatem portum futurum intellegebant. At quem ad modum est pervagatus! Radices palmarum agrestium, quas in nostris navibus invenerant, iactabant, ut omnes istius improbitatem 25 et calamitatem Siciliae possent cognoscere. Siculosne milites, aratorumne liberos, quorum patres tantum labore suo frumenti exarabant ut populo Romano totique Italiae suppeditare possent, eosne in insula Cereris natos, ubi primum fruges inventae 30 esse dicuntur, eo cibo esse usos a quo maiores eorum ceteros quoque frugibus inventis removerunt! Te praetore Siculi milites palmarum stirpibus, piratae Siculo frumento alebantur! O spectaculum miserum atque acerbum! ludibrio esse urbis gloriam, populi 35 Romani nomen, hominum honestissimorum conventum atque multitudinem piratico myoparoni! in portu Syracusano de classe populi Romani triumphum agere piratam, cum praetoris inertissimi nequissimique oculos praedonum remi respergerent! 40

§ 32

To cover his tracks Verres decided to execute the ships' captains, without giving them the chance to defend themselves.

Homo certior fit agi nihil in foro et conventu toto die nisi hoc, quaeri ex nauarchis quem ad modum classis sit amissa; illos respondere et docere unum quemque, missione remigum, fame reliquorum, Cleomenis timore et fuga. Quod posteaquam iste 5 cognovit, hanc rationem habere coepit: si hoc crimen extenuari vellet,—nam omnino tolli posse non arbitrabatur,—nauarchos omnes, testes sui sceleris, vita esse privandos.

10 Haec posteaquam acta et constituta sunt, procedit iste repente e praetorio inflammatus scelere furore crudelitate; in forum venit, nauarchos vocari iubet. Qui nihil metuerent, nihil suspicarentur, statim accurrunt. Iste hominibus miseris inno-
15 centibus inici catenas imperat. Implorare illi fidem praetoris, et quare id faceret rogare. Tum iste hoc causae dicit, quod classem praedonibus prodidissent. Fit clamor et admiratio populi tantam esse in homine impudentiam atque audaciam ut aut aliis
20 causam calamitatis attribueret quae omnis propter avaritiam ipsius accidisset, aut, cum ipse praedonum socius arbitraretur, aliis proditionis crimen inferret; deinde hoc quinto decimo die crimen esse natum postquam classis esset amissa.

25 Quid erat autem quod quisquam diceret aut defenderet? "Cleomenem nominare non licet." At causa cogit. "Moriere, si appellaris"; numquam enim iste cuiquam est mediocriter minatus. At remiges non erant. "Praetorem tu accuses? frange
30 cervices." Si neque praetorem neque praetoris aemulum appellari licebit, cum in his duobus tota causa sit, quid futurum est?

§ 33

Their friends and relations had to pay to visit them, to provide them with food, to ensure them a merciful death and even to bury them.

Veniunt Syracusas parentes propinquique miserorum adulescentium hoc repentino calamitatis suae commoti nuntio; vinctos aspiciunt catenis liberos suos, cum istius avaritiae poenam collo et cervicibus suis sustinerent; adsunt, defendunt, proclamant, fidem 5 tuam, quae nusquam erat neque umquam fuerat, implorant.

Includuntur in carcerem condemnati; supplicium constituitur in illos, sumitur de miseris parentibus nauarchorum; prohibentur adire ad filios, prohi- 10 bentur liberis suis cibum vestitumque ferre. Aderat ianitor carceris, carnifex praetoris, mors terrorque sociorum et civium Romanorum, lictor Sextius, cui ex omni gemitu doloreque certa merces comparabatur. "Ut adeas, tantum dabis, ut cibum tibi intro 15 ferre liceat, tantum." Nemo recusabat. "Quid? ut uno ictu securis adferam mortem filio tuo, quid dabis?" Etiam ob hanc causam pecunia lictori dabatur. O magnum atque intolerandum dolorem! o gravem acerbamque fortunam! Non vitam 20 liberum, sed mortis celeritatem pretio redimere cogebantur parentes. Estne aliquid ultra quo crudelitas progredi possit? Reperietur; nam illorum, cum erunt securi percussi ac necati, corpora feris obicientur. Hoc si luctuosum est parentibus, 25 rediman pretio sepeliendi potestatem.

§ 34

You thought to destroy all evidence against you; but I still have witnesses to establish the horrid truth.

Quibus omnibus rebus actis atque decisis producuntur e carcere, deligantur. Feriuntur securi. Laetaris tu in omnium gemitu et triumphas; testes avaritiae tuae gaudes esse sublatos. Errabas,
5 Verres, et vehementer errabas, cum te maculas furtorum et flagitiorum tuorum sociorum innocentium sanguine eluere arbitrabare; praeceps amentia ferebare, qui te existimares avaritiae vulnera crudelitatis remediis posse sanare. Etenim
10 quamquam illi sunt mortui sceleris tui testes, tamen eorum propinqui neque tibi neque illis desunt, tamen ex ipso illo numero nauarchorum aliqui vivunt et adsunt, quos, ut mihi videtur, ad illorum innocentium poenas fortuna et ad hanc causam
15 reservavit. Per deos immortales! quo tandem animo sedetis, iudices, aut haec quem ad modum auditis? Utrum ego desipio et plus quam satis est doleo tanta calamitate miseriaque sociorum, an vos quoque hic acerbissimus innocentium cruciatus et
20 maeror pari sensu doloris adficit? Ego enim cum Herbitensem, cum Heracliensem securi percussum esse dico, versatur mihi ante oculos indignitas calamitatis.

Peroration

§ 35

How does Verres dare to appear in court? The fact is that he has been driven mad by the memory of his crimes and the ghosts of his victims. Heaven and earth cry for vengeance upon such a monster, and will take it if you let him go—but you cannot.

Tametsi de absolutione istius neque ipse iam sperat nec populus Romanus metuit, de impudentia singulari, quod adest, quod respondet, sunt qui mirentur. Mihi pro cetera eius audacia atque amentia ne hoc quidem mirandum videtur; multa 5 enim et in deos et in homines impie nefarieque commisit, quorum scelerum Poenis agitatur et a mente consilioque deducitur. Agunt eum praecipitem Poenae civium Romanorum, quos partim securi percussit, partim in vinculis necavit, partim 10 implorantes iura libertatis et civitatis in crucem sustulit. Rapiunt eum ad supplicium di patrii, quod iste inventus est qui e complexu parentum abreptos filios ad necem duceret, et parentes pretium pro sepultura liberum posceret. Religiones 15 vero caerimoniaeque omnium sacrorum fanorumque violatae, simulacraque deorum, quae non modo ex suis templis ablata sunt sed etiam iacent in tenebris ab isto retrusa atque abdita, consistere eius animum sine furore atque amentia non sinunt. 20 Neque iste mihi videtur se ad damnationem solum offerre, neque hoc avaritiae supplicio communi, qui

se tot sceleribus obstrinxerit, contentus esse: singularem quandam poenam istius immanis atque
25 importuna natura desiderat. Non id solum quaeritur ut isto damnato bona restituantur iis quibus erepta sunt, sed et religiones deorum immortalium expiandae et civium Romanorum cruciatus multorumque innocentium sanguis istius supplicio luendus
30 est. Non enim furem sed ereptorem, non sacrilegum sed hostem sacrorum religionumque, non sicarium sed crudelissimum carnificem civium sociorumque in vestrum iudicium adduximus, ut ego hunc unum eius modi reum post hominum memoriam fuisse
35 arbitrer cui damnari expediret.

Nam quis hoc non intellegit, istum absolutum dis hominibusque invitis tamen ex manibus populi Romani eripi nullo modo posse? Quis hoc non perspicit, praeclare nobiscum actum iri si populus
40 Romanus istius unius supplicio contentus fuerit, ac non sic statuerit, non istum maius in sese scelus concepisse,—cum fana spoliarit, cum tot homines innocentes necarit, cum cives Romanos morte, cruciatu, cruce adfecerit, cum praedonum duces
45 accepta pecunia dimiserit,—quam eos, si qui istum tot tantis tam nefariis sceleribus coopertum iurati sententia sua liberarint? Non est, non est in hoc homine cuiquam peccandi locus, iudices; non is est reus, non id tempus, non id consilium, (metuo ne
50 quid adrogantius apud tales viros videar dicere,) ne actor quidem est is cui reus tam nocens, tam perditus, tam convictus aut occulte subripi aut impune eripi possit.

§ 36

I call to witness all the gods whose temples and rites he has profaned. I have done my duty. Do you, gentlemen of the jury, do yours; and may I never have to plead such another case.

Nunc te appello, Iuppiter Optime Maxime, cuius iste donum regale, dignum tuo pulcherrimo templo, dignum Capitolio atque ista arce omnium nationum, dignum regio munere, tibi factum ab regibus, tibi dicatum atque promissum, per nefarium scelus de 5 manibus regiis extorsit; teque, Minerva, quam item duobus in clarissimis et religiosissimis templis expilavit, Athenis, cum auri grande pondus, Syracusis, cum omnia praeter tectum et parietes abstulit; teque, Mercuri, quem Verres in domo et in privata 10 aliqua palaestra posuit, P. Africanus in urbe sociorum. Vosque etiam atque etiam imploro et appello, sanctissimae deae, quae illos Hennenses lacus lucosque incolitis, cunctaeque Siciliae, quae mihi defendenda tradita est, praesidetis, a quibus inventis 15 frugibus et in orbem terrarum distributis omnes gentes ac nationes vestri religione numinis continentur; ceteros item deos deasque omnes imploro et obtestor, quorum templis et religionibus iste nefario quodam furore et audacia instinctus bellum 20 sacrilegum semper impiumque habuit indictum, ut, si in hoc reo atque in hac causa omnia mea consilia ad salutem sociorum, dignitatem rei publicae, fidem meam spectaverunt, si nullam ad rem nisi ad officium et virtutem omnes meae curae vigiliae 25 cogitationesque elaborarunt, quae mea mens in

suscipienda causa fuit, fides in agenda, eadem vestra sit in iudicanda; deinde uti C. Verrem, si eius omnia sunt inaudita et singularia facinora
30 sceleris, audaciae, perfidiae, avaritiae, crudelitatis, dignus exitus eius modi vita atque factis vestro iudicio consequatur, utique res publica meaque fides una hac accusatione mea contenta sit, mihique posthac bonos potius defendere liceat quam im-
35 probos accusare necesse sit.

NOTES ON THE TEXT

§ 1

l. 2. **aliquo modo aliquando**] Note the emphasis of the two related words. Tr. "at last, as best I may". For the obstructions put in Cicero's way, see Introduction, pp. 27–8.

l. 4. **ea...provincia**] The separation of the adjective from its noun gives it greater prominence. In English we should make it the predicate: "That is the province which...."

l. 8. **cum...tum praecipue**] *cum...tum* are often used to signify "not only...but also". Here the addition of *praecipue* makes the contrast more marked. Tr. "though ...yet especially".

l. 9. **rationem...habere**] a metaphor from accounts.

l. 12. **nationum exterarum**] Cicero uses "exterae nationes" in the sense of (i) states paying tribute to Rome, (ii) states wholly outside the "imperium Romanum". Here tr. "foreign states".

l. 13. **prima omnium provincia**] i.e. 241 B.C.

l. 18. **venissent**] The subjunctive is generic ("such as once embraced our friendship").

l. 21. **tantae**] "so great", i.e. vast wealth.

l. 26. **hoc** is explained by the accusative and infinitive construction which follows.

l. 30. **ita...ut**] The introduction of *ita* sometimes makes the consecutive clause which follows restrict or limit the meaning of the main clause. The translation will depend on the context. Here tr. "on the understanding that...".

l. 31. **si cuiquam generi**] *cuiquam* suggests universal application ("if to any class whatever").

l. 36. **ut**] repeated from l. 30 for the sake of clearness.

§ 2

l. 2. **antequam proficisceretur**] The subjunctive with *antequam* usually implies purpose or expectancy. But it must always be remembered that the subjunctive in Latin is a much more "delicate" tense than the indicative and can represent finer shades of meaning (e.g. the thoughts of some person other than the writer). So here by using the subjunctive Cicero may be giving us the thoughts or plans of Verres.

l. 7. O praeclare...sermonis] *praeclare* is an adverb modifying *coniectum*. Verres (swine) was a name which boded ill for Sicily, and, as Cicero indicates, the popular forecasts were destined to be only too true.

l. 8. cum...augurabantur] a purely temporal clause modifying *coniectum* ("at a time when men were predicting...").

l. 11. posset] deliberative subjunctive.

l. 11. in quaestura fugam et furtum] In 82 B.C. Verres held his first appointment to one of the higher offices of state as *quaestor provincialis* to the consul Gnaeus Carbo, one of the Marian leaders. While they were on their way to Cisalpine Gaul he deserted Carbo in favour of Sulla and took with him the military chest of which, as paymaster to the troops, he was in charge.

l. 12. in legatione] The *legati* were deputies whom a provincial governor took with him as senior officers of the army in the province. In 80 B.C. Verres went in this capacity to Cilicia under the praetor Gnaeus Dolabella.

l. 14. latrocinia praeturae] Verres had been *praetor urbanus* in 74 B.C.

l. 14. in quarto actu] the three years of his praetorship in Sicily, 73–71 B.C.

l. 20. possit] Note the present subjunctive after secondary tenses to denote the present state of affairs.

l. 21. aliqua ex parte] "even in some measure".

l. 23. suas leges] their own laws before the Roman occupation, which they were allowed to retain.

l. 23. nostra senatus consulta] privileges granted by decree of the Senate.

l. 24. communia iura] rights to which all human beings are entitled.

l. 27. nisi ad nutum istius] "except in accordance with his pleasure".

l. 29. ab eo] referring to *cuiusquam*.

l. 30. Innumerabiles pecuniae...] Cicero in the course of his speech drives home these general charges by reference to particular cases.

l. 34. iudicio] ablative of instrument.

l. 39. classes] to deal with the pirates.

l. 42. regum] such as Agathocles and Hiero, tyrants of Syracuse.

l. 43. ornamento] predicative dative.

l. 45. dederunt] If Cicero is thinking of Marcus Marcellus, this is a rhetorical exaggeration. Livy has a different story to tell: see Introduction, pp. 19–20. After the sack of Syracuse in 212 B.C. Marcellus spared little more than the public buildings.

l. 46. **reddiderunt**] Cicero elsewhere recalls how Publius Scipio Africanus after the capture of Carthage in 146 B.C. restored various works of art which had been taken by the Carthaginians, e.g. a statue of Mercury to Tyndaris (see § 36, l. 11).

l. 47. **At enim**] used to introduce a possible objection; tr. "But, some one may say...."

l. 52. **ne qua**] After *si*, *nisi*, *num*, *ne*, "any" is translated by the indefinite *quis* (adjective *qui*).

l. 54. **causam**] "the facts of the case".

§ 3

l. 1. **rerum capitalium quaestionibus**] "criminal actions", i.e. trials involving the *caput* (life or civil status) of the accused.

l. 5. **Halicyensis**] a native of Halicyae in Sicily, one of the five *civitates immunes ac liberae*. See Introduction, p. 21. Such states were not generally amenable to the Roman jurisdiction, but in criminal cases the governor did occasionally intervene. It is also possible that the alleged offence occurred outside Halicyae in a place which did not enjoy the privileges of immunity.

l. 5. **cum primis**] among the first, i.e. "especially". So also *in primis*, l. 17.

l. 6. **C. Sacerdotem**] Verres' predecessor in Sicily.

l. 8. **eo iudicio**] "as a result of the verdict given on the case".

l. 8. **Sopatro**] Dative of disadvantage.

l. 10. **nomen detulerunt**] Supply *ad praetorem*; a legal phrase meaning "they prosecuted".

l. 13. **Citatur reus**] The historic present and short sentences are used for the sake of vividness. *causam agere* = to try a case.

l. 20. **videretur**] The subjunctive is generic.

l. 26. **ob salutem**] "to grant an acquittal".

l. 26. **si fieri posset**] The historic present takes either a primary or a secondary sequence, sometimes both.

l. 28. **illi** = *Sopatro*.

l. 35. **hominem ad HS LXXX perducit**] HS = IIS (*duo et semis*) = 2½ asses = 1 sestertius. HS also stood for sestertia, the plural of sestertium which was not a coin but a sum of money equivalent to 1000 sestertii (about £8). Hence HS LXXX might equal 80 or 80 thousand sesterces; but to avoid ambiguity the larger number is usually written HS $\overline{\text{LXXX}}$ (*sestertia octogena*). HS $\overline{\text{LXXX}}$ (*sestertium octogiens*) would represent 8 million sesterces. Tr. "he induced the fellow to accept 80,000 sesterces" (about £650).

l. 36. **numerat**] suggests he paid cash down!

§ 4

l. 9. **videret**] represents an imperative in the direct speech.

l. 11. **et iure iniquo**] Because he was a Sicilian and therefore at a disadvantage in a court of law. See footnote, Introduction, p. 14.

l. 11. **et tempore adverso**] Because he was a man on trial and therefore in a difficult position. The ablatives are descriptive (ablatives of quality).

l. 15. **iste**=Verres. Frequently used half-contemptuously of the accused—the man you know.

l. 17. **in consilio**] on the bench as gentlemen of the jury.

l. 19. **tum cum est...absolutus**] The demonstrative *tum* emphasises the purely temporal nature of the *cum* clause. Hence the indicative.

l. 20. **rationis**] partitive genitive dependent on *hoc*. So also *id aetatis*, *aliquid periculi*, etc.

l. 21. **eisdem testibus**] ablative absolute.

l. 23. **hac una spe**] explained by (i) *consilii frequentia et dignitate*, (ii) *quod erant*....

§ 5

l. 10. **rei privatae iudex**] The judge who presided at a private suit between two parties.

l. 10. **recusabat**] Note the tense—"was inclined to refuse".

l. 11. **quos...vellet**] i.e. to advise him. *Vellet* is subjunctive as being part of the words used by Petilius and is virtually a subordinate clause in oratio obliqua, the *quod* clause introducing the statement and giving his grounds for refusal.

l. 14. **adesse**] used in a technical sense—to be present in court with someone (dative) in order to help him.

l. 17. **habebant**] not regarded as part of what they said: therefore indicative.

l. 20. **dimisisset**] See on *vellet*, l. 12. It is all part of what Minucius thought.

l. 21. **cum...iubetur**] The indicative is used with "inverted" *cum*, i.e. when the *cum* clause really contains the principal statement and what would otherwise have been the *cum* clause is made the main clause and put first.

l. 24. **Graeculo**] The diminutive is contemptuous—"a miserable Greek".

l. 25. **pervellem adessent**] i.e. *utinam adessent*, expressing an unfulfilled wish in present time. It is perhaps easiest to treat *pervellem* as the apodosis of a present unfulfilled condition of which the protasis is not expressed. *Volo* and its compounds often take the subjunctive without *ut* when the subjects of the two verbs are different. The prefix *per* intensifies the meaning of the word to which it is attached. Tr. "I could wish they were here".

l. 27. **Nam hercule**] an elliptical use: some such words are implied as "I must leave the court too", for which *nam* gives the reason. *hercule* may be the vocative of an old form *Herculus*. We might translate by: "As a matter of fact...."

l. 30. **verbis...prosequitur**] The comparatives *vehementioribus* and *gravius* are used absolutely. Tr. "assailed him with language of considerable violence".

l. 33. **negotiaretur**] Whenever the relative is used with the subjunctive, the clause is adverbial instead of adjectival. A purely adjectival clause has its verb in the subjunctive only if it forms part of an oratio obliqua. Here the sense is causal ("inasmuch as he...").

l. 33. **sic...ut...meminisset**] "without forgetting".

l. 34. **ita...ut ne**] *ut ne* shows that the restrictive clause is to be classed as final rather than consecutive. Tr. "with the precaution that".

l. 35. **rem**=property, wealth.

l. 36. **quae tempus...causa**] "as the particular circumstances and the case required".

§ 6

l. 1. **importunitate et audacia**] ablatives of quality.

l. 3. **ageret...verteret**] indirect deliberatives.

l. 4. **dimisisset**] represents a future perfect in oratio recta. So also *condemnasset* and *rescidisset*.

l. 7. **ita**] explained by the *cum* clause which follows: in the absence of jury and counsel for the defence.

l. 11. **aestuabat**] Note the metaphor: he was tossed to and fro.

l. 19. **Age**] frequently used with another verb in the imperative: Come! quick!

l. 19. **orare atque obsecrare**] historic infinitives.

l. 20. **cognosceret**] here used in a technical sense: examine or try a case.

l. 21. **dicit**] a common word which, like *res*, must be translated to suit the context. Here the sense is "gave evidence". *Interrogatur* suggests cross-examination.

l. 22. **praeco dixisse pronuntiat**] The normal procedure was that, when both sides had stated their case, the herald of the court (*praeco*) said: *Dixerunt*. The jury then considered their verdict. Tr. "the herald announced that the speeches were finished".

l. 25. **ita properans**] "with such haste", i.e. *quasi metueret ne....*

l. 25. **de sella exsilit**] to ask the verdict of the "jury" mentioned in l. 27.

§ 7

l. 4. **Thermitanus**] a citizen of Thermae, on the north coast of Sicily.

l. 10. **paulo magis**] "particularly, exceptionally". So *paulo studiosus*: "with exceptional keenness". The comparatives are absolute.

l. 12. **supellectilem**] vessels for household purposes (e.g. cups), not furniture in our narrower sense of the term.

l. 13. **et Deliacam et Corinthiam**] Bronze (*aes*) is an alloy composed of about nine parts of copper to one of tin. Delian and Corinthian ware, which was especially famous, was an alloy of bronze obtained by mixing other metal. Corinthian bronze, we are told, was first made by accident from the mixing of molten copper, gold and silver, during the burning of the city by the Romans in 146 B.C.

l. 14. **argenti**] "silver vessels, plate". The genitive is partitive, to be taken with *satis*.

l. 18. **ut poterat**] "as best he could".

l. 18. **animi**] a locative, used with certain words expressing doubt or anxiety.

l. 21. **praetoris...hospitis**] subjective genitives ("suffered at the hands of...").

l. 27. **suam...seque**] Notice that the reflexives refer to different persons. However there is no ambiguity.

l. 30. **monumenta P. Africani**] See § 2, l. 45.

l. 34. **ex oppidi nomine et fluminis**] Himera was a city founded in 649 B.C. on the north coast of Sicily near the mouth of the river Himera. After its destruction in 408 B.C. a new city Thermae was founded in the neighbourhood.

l. 35. **Stesichori**] a Greek lyric poet born in Sicily in the seventh century B.C., who lived part of his life at Himera.

l. 37. **tota Graecia**] The ablative with nouns to describe "place where" does not take a preposition if the adjective *totus* is used.

l. 40. **mire**] best taken as an adverb of degree modifying *scite et venuste*.

l. 43. **agereturque ea res**] *agere* is used in this technical sense with the accusative or *de*+ablative: see § 8, l. 16. Tr. "the matter was dealt with, discussed".

l. 51. **adhuc**] Take with *prope solum*—"so far almost the only one".

l. 53. **eius modi rerum**] *rerum* is partitive genitive after *nihil*; *eius modi* is a genitive of quality to be taken closely with *rerum* and equivalent to *talium*.

§ 8

l. 1. **hasce eius cupiditates**] Notice this use of the plural of abstract nouns. Tr. "instances of this avarice of his".

l. 3. **domo eius emigrat...emigrarat**] *emigrat* implies "moving" from one house to another; Cicero corrects this by substituting *exit* ("goes off"). *atque adeo* is used to make a correction. Tr. "he leaves his house with his belongings (*emigrat*) or rather merely leaves it (*exit*); for he had already previously moved off his belongings (i.e. his ill-gotten gains)".

l. 10. **illi**] dative of disadvantage (=*in illum*).

l. 18. **reddidisset**] For change of sequence see note § 3, l. 26.

l. 19. **ut...reiceret**] This *ut* repeats for clearness the *ut* in l. 15.

l. 22. **hora nona**] The period between sunrise and sunset was divided into twelve equal parts (*horae*), as also the period between sunset and sunrise. These "hours", of course, varied in length according to the time of year.

l. 24. **id istum agere ut**] "that his object was...".

l. 27. **id aetatis**=*ea aetate*. The genitive is partitive, the accusative adverbial. Tr. "a man of that age".

l. 37. **Venerios**] Supply *servos*; literally = slaves of Venus, who was worshipped in her temple on Mt Eryx in the north-west of Sicily. They were employed by Verres to summon people before him and generally assist him in his schemes.

l. 40. **de foro non discedit**] In the forum, as the chief place of meeting in a city, was conducted a variety of business, comprising affairs of state, financial transactions and administration of justice. Hence the various phrases in which the word is used, e.g. *cedere foro* = to become bankrupt. Here tr. "did not leave the court".

l. 41. **Agathinum**] one of the *inimicissimi*; see § 8, l. 5.

l. 45. **verbo posuit**] "alleged".

l. 48. **videri**] Not that Verres meant it to appear doubtful, but because *videtur* was regularly used by a judge in pronouncing judgment, on the principle that human judgment is not infallible.

l. 50. **HS D** = HS $\overline{\text{D}}$: 500,000 sesterces.

l. 55. **palam de sella ac tribunali**] See § 11, l. 18.

l. 57. **nomen recepturum**] a technical phrase—receive an accusation. Cf. *nomen deferre*—bring an accusation.

l. 61. **adfinem**] "implicated in" + genitive.

l. 64. **Iste vero**] Some word of saying must be supplied.

§ 9

l. 1. **Hic**] i.e. Sthenius.

l. 7. **placere**] impersonal—"that it was their opinion".

l. 8. **statui**] infinitive dependent on *placere*—"that a decree should be passed (to the effect that...)".

l. 14. **summa voluntas**] a strong inclination to support the motion and counteract Verres (the son).

l. 17. **si quod**] Supply *iudicium*.

l. 19. **id temporis**] i.e. after sunset. The Senate as a rule did not put a motion to the vote after sunset; this explains the obstructive tactics of the elder Verres, mentioned in the next line. Tr. "so late".

l. 20. **qui...consumerent**] The relative with the subjunctive here expresses purpose.

l. 28. **fore ut...revocetur**] a periphrasis necessary with verbs which have no supine and often used elsewhere instead of the future infinitive.

l. 32. **isti...in integro**] i.e. he had as yet taken no steps in the matter. *In integro* = *intacta*; *isti* is a dative of the person affected, here equivalent to a dative of the agent.

l. 35. **in eo, quod monebatur**] "so far as advice was concerned".

l. 37. **voluntatem**] i.e. his better feelings.

l. 41. **nescio quo casu**] "by some chance or other".

l. 45. **ego**] Note the emphasis: "I at any rate".

l. 45. **a Vibone Veliam**] in southern Italy. With towns and small islands *a* is often used in the sense "from the neighbourhood of".

l. 46. **fugitivorum**] runaway slaves. It was in the neighbourhood of Vibo, in the territory of the Bruttii, that Marcus Crassus ended the Slave War in 71 B.C.

l. 46. **praedonum**] A reference to the pirates, who about this time were at the height of their power. See Introduction, pp. 23–4.

l. 47. **tua tela**] Cicero was afraid he might be assassinated by hirelings of Verres.

l. 49. **ex reis eximerere**] "your name should be removed from the list of the accused".

l. 49. **si...adfuissem**] the apodosis is suppressed ("as it would have been if...").

l. 49. **ad diem**] i.e. the 110th day. As a matter of fact Cicero was back by the 50th day, probably because he felt that, in dealing with a man like Verres, Rome was a safer place than Sicily. See Introduction, p. 27.

§ 10

l. 2. **Cognitorem**] "an attorney", to act for Sthenius in his absence.

l. 5. **At**] "at least".

l. 7. **Cui**] dative of agent with *probari*.

l. 9. **amplissima...amicitiis**] The ablatives are descriptive. Tr. "with a wide circle of relations and numerous friends".

l. 10. **multum**] a cognate accusative used very much like an adverb.

l. 12. **qui...fieret**] final.

l. 12. **probabis**] Cicero addresses Verres. Note the difference in meaning between *probabis* here and *probari* in l. 8.

l. 13. **At** = *at enim*. *credo* is used ironically.

l. 14. **eligere**] Supply *Verrem*.

l. 15. **suorum**] i.e. of Verres.

l. 16. **in tabulis**] "in the official records".

l. 17. **Palatina**] ablative=*ex tribu Palatina*. The Palatina tribus was one of the four city tribes, named after one of the seven hills of Rome.

l. 20. **secutus est**] "was attracted, influenced by".

l. 21. **Quid?**] Some verb (e.g. *ais*) must be supplied if we insist on a grammatical explanation. Tr. "Tell me this!"

l. 22. **cum...tum**] The emphasis is on the words which follow *tum*.

l. 24. **contra venit**] a technical expression—"appeared against him".

l. 25. **utrum...an**] Either alternative was damaging to Verres.

l. 26. **ad Sthenii periculum**] a special legal meaning of *periculum*. Tr. "in the record of his sentence".

l. 31. **subcrispo**] Notice the force of the prefix here and in *peracutus*, l. 32.

l. 31. **nigrum**] i.e. his complexion was dark.

l. 33. **monet**] i.e. "prompting him".

l. 34. **proximus**] i.e. *Verri*.

l. 35. **sequester...interpres**] In cases of bribery and corruption the *sequester* was the man who received the money from the briber and afterwards paid it to the bribed; the *interpres* was the go-between who arranged the business between the parties.

l. 36. **Timarchidi**] See § 3, l. 21. Note the form of the genitive.

l. 37. **Apronio illi**] "The notorious Apronius" was an agent of Verres, elsewhere described as chief of the *decumani*, who exacted the tenths (*decumae*). Cicero gives him a very bad character. Note the use of *ille* of someone who is famous or infamous.

l. 41. **cui...imponeret**] Here the sense is final; with *quem...arbitraretur* in the following clause it is generic.

§ 11

l. 1. **primis illis temporibus**] i.e. after the condemnation of Sthenius in his absence. Ablative of time within which.

l. 5. **vertit stilum**] i.e. he rubbed out what he had written. The sharp end of the pen (*stilus*) was used to write on the wax tablets (*tabulae*); the reverse end was flat and could be used to smooth the wax and erase what had been written.

l. 6. **evertit**] Cicero was very fond of puns.

l. 15. **facit coram esse delatum**] Supply *nomen Sthenii*. Tr. "makes an entry to the effect that the accusation was brought against him in his presence".

l. 18. **palam**] i.e. in the forum, where all could hear.

l. 18. **de loco superiore** = *de tribunali*, the raised platform on which his magistrate's seat (*sella*) would be placed.

l. 22. **hoc**] ablative; explained by what follows.

l. 30. **Cedo**] probably the second person singular of an old imperative. Tr. "Give me".

l. 37. **Ex istis...rebus**] Note the order of words and the emphasis.

l. 40. **mihi**] ethic dative.

§ 12

l. 4. **vos...penditote**] Note the alliteration. Tr. "You can estimate it at its own value irrespective of the name."

l. 5. **Genus ipsum**] i.e. *criminis*: the nature of the charge.

l. 8. **in Sicilia tota**] The preposition is here used for emphasis.

l. 9. **oppidis...familiis**] ablatives of quality, describing *provincia*.

l. 10. **Corinthium aut Deliacum**] For Corinthian and Delian bronze ware see § 7, l. 13.

l. 14. **neque in tabula neque in textili**] Verres was only interested in paintings which could be easily removed. *textili* suggests embroidered or woven work, not painting on canvas, which was not introduced till later. Tr. "whether on panels or tapestry".

l. 17. **Latine** = *plane*: in plain Latin!

l. 19. **locis communibus**] "public places".

l. 21. **istum**] part of the accusative and infinitive with *reliquisse*.

l. 23. **neque...neque**] Negatives may be repeated, without producing an affirmative, when the first negative (*nihil*) is general. Cf. *ne...quidem*, ll. 19, 20.

§ 13

l. 1. **Mamertinus**] a citizen of Messana in Sicily, the modern Messina on the strait separating the island from the mainland of Italy.

l. 1. **illa in civitate]** Messana.

l. 2. **vel]** "perhaps"; used in a restrictive sense. The meaning is: the best known, at any rate (*notissima quidem certe*), if not the finest.

l. 11. **idiotas]** *idiota* = the Greek ἰδιώτης, which, from being originally used of an individual in a private as opposed to a public capacity, came to be applied to a person who was ignorant not only of politics, but of any science or art.

l. 13. **Praxiteli]** genitive. Praxiteles was a famous Greek sculptor of the fourth century B.C. Perhaps his greatest work was a statue of Aphrodite.

l. 13. **nimirum didici]** As a matter of fact Cicero, like his friend Atticus, was a connoisseur in these matters, but he was well aware that most of his audience, at any rate, shared the old republican prejudice of Cato against art, especially Greek art.

l. 13. **dum...inquiro]** during his stay in Sicily while he was collecting evidence for the trial.

l. 16. **Thespiae]** a town in Boeotia near Helicon, the mount of the Muses.

l. 21. **Myronis]** born in Boeotia: Myron was a famous artist in bronze who flourished in the fifth century B.C.

l. 22. **arulae]** "little altars"; diminutive from *ara*.

l. 27. **Canephoroe** = the Greek κανηφόροι (basket carriers); they were Athenian maidens who, in religious processions, carried on their heads a basket (κανοῦν), containing offerings. Statues of them were used in architecture as supporting columns.

l. 28. **quem?...admones]** a rhetorical trick: these words were never spoken. See Introduction, p. 28.

l. 29. **Polyclitum]** another celebrated Greek sculptor, who lived in the fifth century B.C.

l. 35. **tamen]** corrective—"at least, or rather".

l. 38. **Pro...fidem]** *pro* is an interjection; *fidem* is an exclamatory accusative. So also *defensionem*, § **14**, l. **3**. See § 26, l. 29.

l. 41. **imperio]** the military power held by commanders-in-chief, consuls, praetors, and censors. *potestas* was the civil power of magistrates.

l. 49. **praeterea]** besides himself.

l. 51. **superiorum]** "predecessors".

l. 55. **in hereditate sacrorum]** Cicero here uses *sacra* to include not only the statues, but the sacred rites generally, which were handed down in a Roman family from one generation to another.

§ 14

l. 4. **securibus**] an outward and visible sign of his imperium as praetor, in command of a province. A praetor on duty at Rome would not have *secures* carried by the lictors, but only *fasces* (see l. **21**).

l. 6. **coemeret...relinqueret**] final.

l. 10. **ipsum**] "in his own handwriting".

l. 10. **in tabulas referre**] "to enter in his account books".

l. 12. **HS sex milibus quingentis**] about £52.

l. 15. **Cupidinem...HS MDC**] about £13. The construction must be explained by the ellipse of some words like *te emere* (exclamatory accusative and infinitive).

l. 18. **aliqua**] for *ulla*, because *neque* is regarded as going more closely with the verb.

l. 21. **fascibus**] here in the sense of "authority as praetor". *Fasces* were rods carried before the highest magistrates by the lictors who attended them.

l. 23. **eripuisse**] i.e. *haec signa*.

§ 15

l. 2. **Cibyratae**] from Cibyra, a town in Phrygia (Asia Minor).

l. 4. **fingere e cera**] "to model in wax".

l. 9. **ad eum...contulerunt**] when he was in Asia serving as legatus to Gnaeus Dolabella in 80 and 79 B.C.

l. 12. **canes...diceres**] *diceres* is the indefinite 2nd person singular (cf. French *on*): it is the apodosis of an unfulfilled condition, of which the protasis is easily understood ("if you had seen them"). Tr. "you would say they were bloodhounds".

l. 15. **aliud...aliud**] Note the change of order for variety. Cicero begins by placing the pairs in the same order (*a b*, *a b*) and then reverses the order (*a b*, *b a*), making use of chiasmus, by which contrasting words are arranged crosswise (like the Greek letter χ).

l. 18. **argentum**] silver plate.

§ 16

l. 1. **Verum**] adjective, agreeing with *hoc*, used predicatively.

l. 1. **mehercule**] equivalent to (*ita*) *me Hercule* (*iuves*). Tr. "on my honour".

l. 2. **Lilybitanum]** a citizen of Lilybaeum, on the south-west coast of Sicily.

l. 3. **Boethi]** a Carthaginian artist who did work in relief.

l. 10. **Venerius]** See note, § 8, l. 37.

l. 10. **sigillatos]** with *sigilla* on them (diminutive of *signum*), i.e. little figures.

l. 12. **binos**=a pair. *Duos* would imply two, not necessarily a pair.

l. 12. **utrosque**=both of them.

l. 20. **ut...ne**=*ne*.

l. 21. **Ne multa]** Supply *dicam*. "In short".

l. 21. **HS mille]** about £8.

l. 23. **coepisse]** historic infinitive.

l. 25. **luteum negotium]** *negotium* (for *res*) is used colloquially. Tr. "paltry stuff".

l. 28. **hoc**=*ista intellegere*, i.e. such knowledge of works of art as Verres had.

l. 29. **nugatorium]** See § 13, l. 13.

l. 32. **simile hominis**=*humanitatis* (culture).

l. 32. **Tum primum]** contrasted with *antea*, l. 28.

§ 17

l. 1. **ideo]** inasmuch as *leviora videbuntur*.

l. 2. **quod...Syracusis]** Cicero proceeds to enumerate what he professes to pass over—a well-known rhetorical trick.

l. 2. **mensas Delphicas]** Tables supported on three legs, like the Delphic tripod. Besides their use in temples they served as sideboards in private houses.

l. 7. **mystagogos]** a Greek word, used of those who initiated into mysteries. Here=guides (cf. dragomans).

l. 18. **reges Syriae]** Actually Tigranes of Armenia was in possession of Syria at this time, having expelled their father in 83 B.C. The two sons, their father now dead, came to Rome in 75 B.C. to press their claim to Egypt, in company with their mother, an Egyptian princess. They were unsuccessful and returned in 73 B.C., but on the defeat of Tigranes by Lucullus in 69 B.C. the elder prince Antiochus was restored to the kingdom of Syria until Pompey made it a Roman province in 65 B.C.

l. 20. **temporibus]** difficult times, e.g. owing to the wars against the slaves and pirates.

l. 29. **Mittit homini]** *ad*+accusative would be more

usual, but the dative is used when some other idea than that of motion predominates (here = *dat*).

l. 35. **Quid multa?**] Supply *dicam*. "To be brief".

l. 42. **trulla** = a small wine-ladle; in apposition to *vas* (= an implement, utensil).

l. 42. **manubrio**] ablative of description with *trulla*.

l. 49. **pulcherrima**] Note the Latin order: the superlative is transferred to the relative clause and put in agreement with the relative.

l. 51. **nosset**] The relative clause is causal.

§ 18

l. 4. **nondum perfectum**] The temple of Jupiter on the Capitoline hill was said to have first been built by the Tarquins. It was burnt down in 83 B.C.; the new temple was dedicated in 69 B.C.

l. 15. **animo et puerili esset et regio**] ablatives of description (quality); two reasons why he granted Verres' request (therefore *esset*, not *erat*).

l. 22. **videretur**] *cum* temporal + subjunctive always has the secondary sequence, even with the historic present. So also *recusaret*, § 20, l. 21.

l. 23. **etiam atque etiam**] "over and over again".

l. 27. **metuere**] As often, historic infinitives, present tenses, short sentences and absence of connecting words, denote the graphic or conversational tone of a passage.

l. 28. **si videatur** = "if convenient", in contrast to *Iubet iste*.

l. 30. **reverti**] i.e. *eos* (the servants of Antiochus).

l. 31. **Ipse hominem appellat**] *appellare* is used of a creditor addressing a debtor. Tr. "he applies to the fellow in person".

l. 37. **Iovis**] objective genitive, whereas *hominum* is subjective. *Religio* here means the religious scruples felt by a man who had virtually consecrated the candelabrum to Jupiter; see l. 53.

§ 19

l. 5. **regibus**] in apposition to *maioribus*.

l. 6. **praeceps provincia populi**] Note the effect of the alliteration of the (ex)plosive letter *p*.

l. 7. **nationes exteras**] "foreigners". See § 1, l. 12.

l. 11. **odio atque acerbitati**] predicative datives.

§ 20

l. 6. **Tyndaritanis**] the people of Tyndaris, on the north coast of Sicily.

l. 10. **anniversariis**] ablative of time at which. *summa religione*: ablative of manner.

l. 16. **ita**] referring, not to the *ut* clause which follows, but to the *tamquam* clauses which precede.

l. 19. **primo illo adventu**] contrasted with *aliquanto post venit*, l. 24.

l. 26. **capitis**] See § 3, l. 1.

l. 27. **si...quisquam**] more emphatic than *si quis*. See § 1, l. 31. Tr. "if anyone at all".

l. 28. **Quam mihi...narras**] A colloquial usage. Note the emphatic position of *mihi*.

l. 31. **traditur**] Note the tense ("unless you give it up now").

l. 32. **rem defert**] Cf. *refert*, l. 22. The difference, if any, consists in referring a matter (*referre*) and merely informing (*deferre*).

l. 38. **in conventu palam**] The court would be held in the forum.

§ 21

l. 3. **de porticu**] Verres himself was "under cover".

l. 6. **lictoribus**] ablative of instrument used instead of ablative of agent (especially in military phrases).

l. 7. **videres**] See § 15, l. 12.

l. 7. **id**=*ut...caederetur*.

l. 8. **fefellit...opinio**] "this is where they were mistaken".

l. 10. **Marcellorum**] M. Marcellus, conqueror of Sicily, and his descendants. The family were patrons of Sicily.

l. 12. **C. Marcelli**] great-grandson of M. Marcellus: he went as governor to Sicily in 79 B.C. and was one of the judges at this trial.

l. 18. **in aere**] "on bronze" (considerably colder than wood or stone).

l. 19. **iniuriae crudelitatique**] hendiadys ("one by two"), whereby two nouns are used instead of one noun with an adjective in agreement. Tr. "cruel outrage".

§ 22

l. **4. huius generis**] referring to the plundering of statues, etc.

l. **5. quem ad modum**] but see note, § **2**, l. **45**.

l. **7. annalibus**] histories of Rome.

l. **8. Conferte**] The whole passage is a good example of antithesis. The imperative stands for the protasis of which *dicetis* in l. **13** is the apodosis.

l. **12. constitutas** = *bene constitutas*.

l. **13. iam illa omitto**] See note, § **17**, l. **2**.

l. **15. quod...redundasse**] Cicero is more concerned to point a vivid contrast than to state the facts with due regard to truth.

l. **16. servatum est**] We might have expected the subjunctive—see *clausus fuisset*, l. **19**.

l. **18. tum**] not *introitu Marcelli*, but in the days when Syracuse was a free city.

l. **20. Cilicum myoparoni praedonibusque**] See Part III.

§ 23

l. **1. Insula**] i.e. Ortygia, the original city, separated from the mainland by a narrow channel. See § **31**, ll. **18–22**.

l. **6. Agathocli regis**] tyrant of Syracuse 317–289 B.C.

l. **7. in tabulis picta**] probably on panels, cf. note, § **12**, l. **14**.

l. **11. victoria illa sua**] the sack of Syracuse in **212** B.C

l. **11. profana**] i.e. no longer *sacra religiosaque* (l. **14**), because they had passed into the hands of the victor and were spoils of war.

l. **15. tot saecula**] accusative of "time during which".

l. **22. argumenta**] "subjects".

l. **23. Gorgonis**] i.e. the Gorgon, Medusa.

l. **25. et tamen**] "all the same"—explained by *nam...*, l. **26**.

l. **26. bullas**] studs, knobs, for purposes of ornament.

l. **37. rationem redditurus**] "render an account".

l. **39. quo...abstulisset**] See note on the subjunctive § **2**, l. **2**.

l. **39. quo...eo**] used with comparatives, as correlatives.

l. **40. minore periculo**] ablative of manner. He would have more money for purposes of bribery.

l. **42. de loco superiore**] See § **11**, l. **18**.

l. 43. **ageret**] The relative clause is causal ("inasmuch as he did...").

l. 43. **pro**] "in virtue of". Verres as praetor in Sicily held both *imperium* and *potestas*. See note, § **13**, l. **41**.

§ 24

l. **8**. **Ita...**] explained by the accusative and infinitive following. The meaning is: "The defence are relying on this argument, that...."

l. **11**. **a belli periculis**] the possibility of the slaves in Sicily joining the slaves and gladiators who rose in Italy and were finally put down in 71 B.C.

l. **12**. **quo...conferam**] literally "to what part am I to apply the method of my accusation?" i.e. where am I to direct my attack? *accusationis meae rationem* is little more than a periphrasis for *accusationem meam*.

l. **14**. **quidam**] The literally-minded Romans hesitated to use a strong metaphor, and, when they did so, liked to soften it by the addition of some word such as *quidam* or *quasi* to indicate a figure of speech.

l. **15**. **locum** = "the ground of proof, argument".

l. **15**. **se iactaturus sit**] This *locus* will give Hortensius plenty of scope. Tr. "make a display".

l. **16**. **Hortensius**] See Introduction, p. 27.

l. **16**. **tempora rei publicae**] During Verres' praetorship Rome was struggling with the slaves, the pirates, Sertorius and Mithridates.

l. **18**. **pro suo iure**] "in virtue of his right", possibly as leading barrister of the day.

l. **20**. **Romano Siculorum**] Note the contrast effectively brought out by the order of words.

l. **25**. **Sit**] concessive. The defence are prepared to admit this.

l. **27**. **at**] emphatic—"but, they say,...".

l. **29**. **mihi...subici**] to have it brought to my notice, i.e. "to be prompted".

l. **31**. **Habetis** = "you are well aware of".

l. **33**. **Summa**] a noun. "The main thing is this...."

l. **33**. **ut sciatis**] explanatory of *illuc*.

l. **35**. **ne qui**] for *ne quis*.

§ 25

l. **1**. **vel**] as often with superlatives—in the highest degree.

l. **13**. **tectum...lectum**] note the play on words.

l. 14. **viderit**] In consecutive clauses the perfect subjunctive is frequently used after secondary tenses. This tense is aoristic, i.e. it expresses a definite and completed past event.

l. 16. **Cum...coeperat**] When *cum* = *quoties* (whenever) it is regularly used with the pluperfect indicative in past time. So *cum rosam viderat*, l. 18.

l. 17. **a Favonio**] The west wind, which blew at the beginning of spring.

l. 17. **aliquo astro**] Some particular star; *ullo* would be universal (any).

l. 18. **rosam**] i.e. Verres waited as long as he could, in fact till the roses were in bloom (months earlier in Sicily than in England). This seems cynical. Villainy is not inconsistent with a fondness for roses.

l. 21. **ut mos...regibus**] He travelled about like an Eastern potentate.

l. 22. **lectica**] a sort of palanquin adopted at Rome from the East.

l. 22. **octaphoro**] in apposition to *lectica*. It means a *lectica* carried by eight bearers.

l. 23. **Melitensis**] made of Maltese linen, for which Malta was famous.

l. 24. **rosa fartus**] "stuffed with rose leaves".

l. 24. **coronam**] Chaplets were frequently worn on convivial occasions: Verres seems to have extended the use.

l. 26. **minutis maculis**] "of fine mesh".

l. 33. **iura discripserat**] i.e. *singula singulis*; each individual was treated separately. The ablatives *pretio* and *aequitate* are instrumental.

l. 34. **Libero**] Liber was identified with the Greek Bacchus.

l. 40. **res...vocabatur**] "matters came to...".

l. 43. **illis legibus**] What Horace calls *leges insanae*, laws of the drinking prescribed by the *arbiter bibendi*, a sort of Master of the Ceremonies, chosen from among the guests by lot before the drinking began, who decided the strength of the wine and the amount each person was to have.

l. 49. **Cannensem pugnam**] Nearly the whole Roman army was annihilated at Cannae by Hannibal and the Carthaginians in 216 B.C.

l. 51. **pallio**] a Greek cloak and therefore not the garb of a manly Roman, especially a governor.

l. 52. **tunica**] an under-garment or sort of shirt. A *tunica talaris* (stretching to the ankles) was also a sign of effeminacy.

l. 54. **non ius dici]** *ius dicere* was a technical term used of the magistrate who exercised his *iurisdictio*. Tr. "judicial decisions were not given".

l. 54. **non iudicia fieri]** "legal proceedings were not held".

l. 57. **causarum atque iuris]** Cf. l. 54.

§ 26

l. 1. **defendis]** "maintain in his defence".

l. 5. **Esto]** used contemptuously to sum up. Tr. "Enough!"

l. 7. **ne quod]** Supply *bellum*.

l. 8. **at vero]** A contrast is made with the preceding sentence and strengthened by *vero*. Tr. "but, on the other hand...".

l. 9. **habuit]** Note the difference of meaning with *classem* and *diligentiam*.

l. 14. **maiestatis]** Supply *laesae* or *minutae*: high treason. *maiestatem minuere* = to lessen the dignity and power of the Roman people or its representatives.

l. 19. **frumento]** The dative goes closely with *sumptum* and comes under the heading of "work contemplated" (cf. *dies colloquio*, "a day for a conference"). It has the same force as a genitive ("cost of the corn for the fleet").

l. 22. **ut]** final.

l. 24. **quaestus]** "opportunities for gain".

l. 25. **accipere]** This and the following infinitives are used as nouns in apposition to *haec* (l. 28).

l. 28. **lucrari]** "put into his own pocket".

l. 29. **Huncine hominem]** exclamatory accusative. *Huncine* is from *hice*, the more emphatic form of *hic*, + the interrogative particle. Such accusatives are best explained as the object of some verb understood.

l. 31. **pro]** "in proportion to".

l. 32. **discribere]** exclamatory infinitive. Also an elliptical construction, i.e. to be explained by the omission of some verb.

l. 33. **missionis]** genitive dependent on *pretium*. Tr. "to fix a definite price, 600 sesterces apiece, for the discharge of sailors!"

§ 27

l. 3. **praetori...praedonibus]** a play on words. Cf. also *ceperunt...captam*, l. 6.

l. 4. **P. Caesetius et P. Tadius**] Verres' quaestor and legatus respectively (cf. l. 10).

l. 5. **navibus suis**] the ablative of accompaniment without *cum* (used in some military and naval phrases).

l. 7. **captam**] not captured, but affected, i.e. encumbered (cf. *oculis captus*: so affected as to be blind).

l. 11. **integra**=*intacta*.

l. 13. **supplicium**] i.e. death.

l. 14. **in hostium numero ducit**] implies that he put them to death.

l. 16. **cohorti**] used of the praetor's suite.

l. 17. **symphoniacos homines**] to be used as household musicians who provided entertainment at banquets.

l. 18. **amico**] dative of advantage.

l. 18. **muneri**] dative of purpose (predicative dative).

l. 21. **sic habent**=*hoc existimant* (explained by the accusative and infinitive following).

l. 23. **humani**=*non indocti*: "well-informed, clever".

l. 26. **securi ferirentur**] "were executed" (by beheading).

l. 27. **ex ipso navigio**] i.e. from its size.

l. 30. **eorum...quos...de piratis**] "of those of the pirates whom...".

l. 40. **in...ad**] Note the uses of the prepositions: *in*=instead of; *ad*=to the amount of.

l. 42. **omnis vestis**] used in a general sense, not only of clothes, but (as here) of any costly coverings such as the pirates may have captured.

l. 43. **aversum**] "misappropriated".

§ 28

l. 9. **regia**] once the property of King Hiero.

l. 10. **velis**] instrumental ablative (of means).

l. 12. **sane**] intensive="very".

§ 29

l. 1. **Centuripina**] of Centuripa, one of the richest towns in Sicily, near the foot of Mt Aetna.

l. 1. **quadriremi**] a vessel having four banks of oars (*quattuor*+*remus*).

l. 2. **Segestana**] For the towns of Sicily see Introduction, p. 21. Segesta was a *civitas immunis ac libera*; the others mentioned here—Tyndaris, Herbita, Heraclia, Apollonia, Haluntium—were *civitates decumanae* (they paid "tenths").

l. 5. **propugnatorum**] soldiers who served on board, marines.

l. 6. **in imperio suo**] Take closely with *classem*; "the fleet under his command".

l. 9. **esset**] The relative clause is concessive.

l. 10. **soleatus**] wearing *soleae* (slippers), generally worn indoors and not in public (when *calcei* would be used).

l. 11. **pallio...tunicaque**] See notes, § 25, ll. 51, 52.

l. 14. **Pachynum**] the south-east promontory of Sicily.

l. 19. **luxurie...nequitia**] ablatives of respect.

l. 23. **portu Odysseae**] a harbour in the vicinity of Pachynus.

l. 26. **erigi...fieri...praecidi**] the accusative and infinitive construction for the more usual *ut*+subjunctive. However, Cicero never uses the active infinitive with *impero*.

l. 29. **velis**] ablative of respect.

l. 31. **moliebantur**] a word implying great effort and strain.

l. 32. **quoquo...habebat**] i.e. in any event.

l. 34. **ferro**] ablative of instrument.

l. 34. **reddere**] "yield, sacrifice".

l. 35. **Quodsi**] "But if".

l. 35. **tanto**] ablative of measure.

l. 36. **ad resistendum**=*resistendi* ("means of resisting").

l. 37. **constrata**="with a deck".

l. 42. **Helorum**] between Syracuse and Pachynus.

l. 61. **cum...invesperasceret**] subjunctive as part of Heracleo's orders.

§ 30

l. 4. **nocte intempesta**] "at dead of night".

l. 12. **dormientem...vigilantem**] "if asleep...if awake".

l. 14. **concursabat**] "was running to and fro".

l. 19. **reliquum**] "in store, impending".

l. 20. **constaret**] impersonal.

l. 21. **ad domum eius**] The preposition is used when *domus*="house" (not "home").

l. 23. **sagum**] a sign of war, as the toga was a sign of peace. So *saga sumere* is a military term often equivalent to "prepare for war".

l. 26. **acta**] i.e. his behaviour on the sea-shore.

l. 30. **ad**] used with names of places in the sense of "at" or "near" (not "in").

l. 31. etiam="still".

l. 45. nomen ac fasces] Verres was the representative of *imperium populi Romani*, and, as such, the *fasces*, symbols of his authority, were carried before him.

§ 31

l. 3. Dicere...disputare] historic infinitives.

l. 10. Qui ex illo numero reliqui] "As many of the survivors who...".

l. 14. testibus] ablative of instrument. See note, § 21, l. 6.

l. 14. tenebatur] "was proved".

l. 16. quo...adisset] =*ut simul atque eo adisset*. *Adisset* is subjunctive because the clause is dependent on *relinqueret* and attracted into the same mood.

l. 18. Insulam] See § 23, l. 1.

l. 19. suo nomine] not "with its own name", but "on its own account".

l. 19. moenibus] i.e. *suis moenibus*—"with walls of its own".

l. 20. maiores] "our ancestors".

l. 26. Siculosne...esse usos] accusative and infinitive of indignant exclamation. Tr. "To think that...." See also § 26, l. 32.

l. 30. in insula Cereris] i.e. sacred to Ceres (Greek Demeter), whose daughter Proserpina (Persephone) was said to have been carried off by Pluto (Hades) while gathering flowers in Sicily. The worship of Demeter was brought to Sicily from Greece.

l. 30. ubi primum fruges...dicuntur] The Sicilians claimed to be the first to receive the gift of corn from Demeter. They held many festivals in honour of Demeter and Persephone, patron goddesses of Sicily. See § 36, ll. 13–18.

l. 38. in portu...piratam] Note the order of words.

§ 32

l. 2. quaeri] impersonal and explanatory of *hoc*.

l. 4. missione] ablative of cause.

l. 14. hominibus] dative after *inici*.

l. 16. hoc causae] The genitive is partitive; the suggestion is that it was no reason at all.

l. 22. arbitraretur] a rare passive meaning.

l. 26. **Cleomenem**] Verres was determined to spare a boon companion.

l. 26. **At**] as frequently, used to introduce an objection (here, the reply of the accused).

l. 28. **mediocriter**] There were no half measures with Verres!

l. 28. **At...erant**] See *remigibus...dimissis*, § **31**, l. **5.**

l. 29. **accuses**] The subjunctive may be explained as the apodosis of a conditional clause, with protasis suppressed. Tr. "Accuse the praetor, would you?"

l. 30. **cervices**] Cicero always uses the word in the plural.

l. 31. **in his duobus**] "in the hands of these two" (Verres and Cleomenes).

§ 33

l. 5. **adsunt**] i.e. to help and defend.

l. 9. **illos** = the sons.

l. 9. **sumitur**] Note the absence of connecting words. The punishment which the parents also had to bear is explained by *prohibentur* etc.

l. 16. **Quid?**] elliptical. See note, § **10**, l. **21.**

l. 26. **redimant**] jussive subjunctive ("let them pay for...").

§ 34

l. 11. **tibi**] to exact vengeance from you.

l. 11. **illis**] to avenge them.

l. 13. **ad...poenas**] i.e. *repetendas*—"to claim vengeance for them".

l. 15. **tandem**] as frequently in questions—"pray, I ask you".

l. 17. **desipio**] "am I weak-minded?"

§ 35

l. 3. **adest**] i.e. *ad iudicium.*

l. 3. **respondet**] "appears"; literally = answers when his name is called by the herald (*praeco*). It is as well to remind ourselves that Verres went into voluntary exile after the Actio Prima and that these words occur in the Actio Secunda which Cicero published after the trial was over.

l. 4. **pro**] "when I bear in mind...".

l. 9. **Poenae**] here personified = the goddesses of Vengeance (Furies).

l. 11. iura libertatis et civitatis] It was illegal to put a Roman citizen to death; and in Cicero's time, at any rate, crucifixion was confined to slaves and aliens (*peregrini*).

l. 34. eius modi] explained by *cui...expediret*.

l. 35. cui damnari expediret] the normal sequence after *fuisse*; *damnari* is the subject of the impersonal *expediret*. Cicero means that the condemnation of Verres is for his good either (i) because it will bring peace to his mind (see ll. 7, 8) or (ii) because if he escapes from this court the Roman People will exact vengeance from him.

l. 36. Nam] This seems to confirm the second reason mentioned above.

l. 36. dis...invitis] ablative absolute ("in defiance of").

l. 39. praeclare nobiscum actum iri] "we shall be very well treated" (i.e. if we acquit him and the Roman People deal with the matter).

l. 40. ac...statuerit] "rather than decide...".

l. 45. si qui] i.e. if there are any who can bring themselves to....

l. 48. peccandi locus] "room for making a mistake".

l. 49. consilium] "panel of jury".

l. 51. actor] "prosecutor".

l. 51. cui] indirect object with *subripi* and *eripi*.

l. 52. subripi] Note the force of *sub*; secretly or in an underhand manner.

§ 36

In this impressive and final summing up Cicero invokes the gods and goddesses insulted and outraged by Verres, at the same time recapitulating the crimes which particularly render him liable to their displeasure.

l. 8. Athenis] when Verres was on his way in 80 B.C. to serve as legatus to Gnaeus Dolabella, the praetor of Cilicia.

l. 8. Syracusis] See §§ 22, 23.

l. 10. Mercuri] See § 20.

l. 11. P. Africanus] See note, § 2, l. 46.

l. 13. sanctissimae deae] Ceres and Libera (an old Italian goddess identified with Proserpina). See note, § 31, l. 30.

l. 13. Hennenses] It was at Henna in the centre of Sicily that Proserpina was carried off by Pluto.

l. 19. religionibus] concrete here—objects of religious veneration, such as statues.

l. 20. **quodam]** *quidam* is used after an adjective to modify it; here the suggestion is that the madness was virtually *nefas*. See note, § 24, l. 14.

l. 27. **vestra]** referring to the *iudices*.

l. 34. **bonos potius defendere]** Cicero rarely appeared in the rôle of prosecutor, and argues elsewhere that in acting in this capacity against Verres he is really defending the Sicilians.

VOCABULARY

ABBREVIATIONS

abl.	ablative.	indecl.	indeclinable.
acc.	accusative.	interr.	interrogative.
adj.	adjective.	irr.	irregular.
adv.	adverb.	m.	masculine.
c.	case, common.	n.	neuter.
comp.	comparative.	plur.	plural.
conj.	conjunction.	prep.	preposition.
dep.	deponent.	pron.	pronoun.
f.	feminine.	1, 2, 3, 4	conjugations.
impers.	impersonal.		

abalieno, 1, *estrange*
abdo, -didi, -ditum, 3, *hide, conceal*
abdūco (dūco), 3, *lead away*
abiūdico, 1, *take away by law*
ablēgo, 1, *send away, remove, discharge*
abripio, -ripui, -reptum, 3, *tear away*
absolūtiō, -ōnis, f. *acquittal*
absolvo, -solvi, -solūtum, 3, *acquit*
absum, abesse, āfui, *am absent*
abūtor, -ūsus, 3, dep. *misuse*
accēdo, -cessi, -cessum, 3, *approach, undertake*
accensus, -i, m. *attendant*
accido, -cidi, 3, *fall upon, reach, happen*
accipio, -cēpi, -ceptum, 3, *accept, take, hear*
accurro (curro), 3, *run to*
accūsātor, -ōris, m. *prosecutor, plaintiff*
accūsātōriē, adv. *in the manner of an accuser*
accūso, 1, *accuse, prosecute*
acerbitās, -ātis, f. *hatred*
acerbus, adj. *severe, bitter*
ācerrimē, adv. *bitterly, sharply*
acta, -ae, f. *sea-shore*
actor, -ōris, m. *advocate*
actus, -ūs, m. *act* (of a play)
adamo (amo), 1, *fall in love with*
addūco (dūco), 3, *bring to, induce*
adeo, adv. *to that extent*
adfero (fero), *bring to*
adficio (facio), 3, *affect, reward, punish*
adfīnis, adj. *neighbouring, implicated in*
adhibeo, 2, *add to*; **in consilium**, *consult*
adhūc, adv. *thus far, hitherto, as yet*
adipiscor, adeptus, 3, dep. *obtain*
adiuvo, -iūvi, -iūtum, 1, *help*
administer, -tri, m. *assistant*
admīrātiō, -ōnis, f. *astonishment*
admitto (mitto), 3, *admit*
admoneo (moneo), 2, *suggest*
adnumero, 1, *pay to*
adorior, -ortus, 4, dep. *attack*
adpello, -puli, -pulsum, 3, *drive forward*
adroganter, adv. *conceitedly, proudly*
adscrībo (scrībo), 3, *appoint*

adsideo, -sēdi, -sessum, 2, *sit by*
adsum, adesse, adfui, *am present*
adulescens, -entis, c. *young person*
adulescentia, -ae, f. *youth*
adveho, -vexi, -vectum, 3, *bring to*
adventus, -ūs, m. *arrival*
adversārius, -ii, m. *opponent, enemy*
adversus, adj. *unfavourable*
advocātus, -i, m. *legal assistant, counsel*
aedis, -is, f. *temple*; (in plur.) *house*
aemulus, -i, m. *rival*
aēneus, adj. *of bronze*
aequus, adj. *honest*
aes, aeris, n. *bronze*
aestās, -ātis, f. *summer*
aestīvus, adj. *summer*; **aestīva**, -ōrum, n. *a summer camp*
aestuo, 1, *am undecided, waver*
aetās, -ātis, f. *age*
agito, 1, *trouble, consider, discuss*
ago, ēgi, actum, 3, *do, make, hold, discuss*
agrestis, adj. *wild*
āio, defective verb, *say*
aliquando, adv. *at some time, at last*
aliquantō, adv. *considerably*
aliquī, adj. *some*
aliquis, pron. *some one*
alius, adj. and pron. *other, another*
alo, alui, altum, or alitum, 3, *feed*
alter, -tera, -terum, pron. and adj. *the one, the other*
alteruter, -tra, -trum, pron. *either, the one, the other*
āmentia, -ae, f. *madness*
amīcitia, -ae, f. *friendship*
āmitto (mitto), 3, *lose*
amoenus, adj. *pleasant*
amplē, adv. *abundantly*
amplus, adj. *great, abundant, much*
ancora, -ae, f. *anchor*
ango, anxi, anctum, 3, *feel distressed*
anguis, -is, c. *snake*
animus, -i, m. *disposition, thoughts, feelings, spirit*
annāles, -ium, m. *historical work, chronicles*
anniversārius, adj. *yearly*
annus, -i, m. *year*
ante, adv. and prep. *before*
anteā, adv. *formerly*
antepōno, -posui, -positum, 3, *place in front, prefer*
antequam, conj. *before*
antīquus, adj. *ancient*
aperio, aperui, apertum, 4, *open*
apertē, adv. *openly*
appello, 1, *call, address, entreat, name*
applico, 1, *apply*; **sē applicāre**, *to devote oneself to*
apud, prep. + acc. *at, before, in the presence of, in the house of*
arātor, -ōris, m. *agriculturist, farmer*
arbiter, -tri, m. *witness, spectator*
arbitrium, -ii, n. *free will*
arbitror, 1, dep. *think*
arcesso, -sīvi, -sītum, 3, *summon*
archipīrāta, -ae, m. *pirate chief*
ardeo, arsi, arsum, 2, *burn*
argenteus, adj. *of silver*
argentum, -i, n. *silver, things made of silver*
argūmentum, -i, n. *evidence, artistic representation*
artifex, -ficis, m. *craftsman*
artificium, -ii, n. *skill*

ārula, -ae, f. *a small altar*
arx, **arcis**, f. *citadel*
aspicio, aspexi, aspectum, 3, *look at*
astrum, -tri, n. *star*
atrōcitās, -ātis, f. *cruelty, harshness*
attendo, attendi, attentum, 3, *consider, listen*
attingo, attigi, attactum, 3, *touch*
attraho, attraxi, attractum, 3, *attract, draw*
attribuo, -ui, -ūtum, 3, *attribute, assign*
auctor, -ōris, m. *adviser*
auctōritās, -ātis, f. *authority, influence*
audācia, -ae, f. *insolence*
audacter, adv. *boldly*
audeo, ausus, 2, semi-dep. *dare*
aufero, -ferre, abstuli, ablātum, 3, *carry away*
augeo, auxi, auctum, 2, *increase, enrich*
auguror, 1, dep. *guess, foretell*
aureus, adj. *golden*
auris, -is, f. *ear*
aurum, -i, n. *gold, things made of gold*
autem, conj. *but*
avāritia, -ae, f. *greed*
avārus, adj. *covetous, greedy*
āverto, -ti, -sum, 3, *steal, convert to my own use*
avītus, adj. *belonging to a grandfather, ancestral*
avus, -i, m. *grandfather*

beneficium, -ii, n. *favour*
benevolentia, -ae, f. *goodwill, kindness, favour*
bīni, adj. *two each, a pair*
bona, -ōrum, n. *goods, property*
breviter, adv. *shortly*
bulla, -ae, f. *boss, knob*

caedēs, -is, f. *slaughter*
caedo, cecīdi, caesum, 3, *cut, beat, kill*
caelātor, -ōris, m. *engraver*
caelum, -i, n. *sky, climate, weather*
caerimōnia, -ae, f. *religious ceremony*
calamitās, -ātis, f. *disaster*
calamitōsus, adj. *disastrous*
calumnia, -ae, f. *trickery, pretence, false accusation*
candēlābrum, -i, n. *candlestick, lamp-stand*
cantus, -ūs, m. *song, singing*
capella, -ae, f. *a she-goat*
capillus, -i, m. *hair*
capitālis, adj. *capital, criminal*
Capitōlium, -ii, n. *the Capitol*
caput, -itis, n. *head*
carbaseus, adj. *of linen*
carcer, -eris, m. *prison*
carnifex, -ficis, m. *butcher, executioner*
cāsus, -ūs, m. *chance, misfortune*
catēna, -ae, f. *chain*
causa, -ae, f. *cause, case*
celeritās, -ātis, f. *speed*
cēna, -ae, f. *dinner*
cēra, -ae, f. *wax*
certus, adj. *trustworthy, settled*
cervix, -īcis, f. *neck*
cēterī, -ōrum, adj. and pron. *the other, the others*
cibus, -i, m. *food*
cingo, cinxi, cinctum, 3, *surround, crown*
circumfero (fero), *carry round*
cito, 1, *summon*
cīvitās, -ātis, f. *state, citizenship*
clam, adv. *secretly*
clāmor, -ōris, m. *shout*
clārus, adj. *celebrated, clear*
classis, -is, f. *fleet*

claudo, clausi, clausum, 3, *shut*
clēmenter, adv. *kindly*
cōdex, -icis, m. *a book*
coemo, -ēmi, -emptum, 3, *buy up*
coepī, coepisse, 3, defective verb, *begin*
cōgitātiō, -ōnis, f. *thought, plan*
cōgito, 1, *think, reflect on*
cognātiō, -ōnis, f. *relationship*
cognātus, -i, m. *relation*
cognitor, -ōris, m. *advocate, attorney*
cognosco, cognōvi, cognitum, 3, *find out, know, try a case*
cōgo, coēgi, coactum, 3, *compel, extort by force*
cohors, cohortis, f. *cohort, retinue, followers*
colligo, collēgi, collectum, 3, *collect*
collum, -i, n. *neck*
colo, colui, cultum, 3, *worship*
commemoro, 1, *remember, relate, mention*
committo (mitto), 3, *commit, entrust*
commoror, -ātus, 1, dep. *remain*
commoveo (moveo), 2, *move, disturb, upset, excite*
commūnis, adj. *common*
comparo, 1, *compare, collect, prepare*
comperio, -peri, -pertum, 4, *find out*
complexus, -ūs, m. *embrace*
complūres, adj. *several, very many*
concēdo, -cessi, -cessum, 3, *yield*
concido, -cidi, 3, *fall*
concipio, -cēpi, -ceptum, 3, *receive, adopt, foster*
conclūdo, -clūsi, -clūsum, 3, *round off*
concupio (cupio), 3, *covet very much*
concurso, 1, *travel about*
concursus, -ūs, m. *running together, assembly*
condo, -ere, -didi, -ditum, 3, *found*
confector, -ōris, m. *a maker, finisher*
confero (fero), *compare, collect, direct*; **sē conferre**, *to betake oneself*
conficio, -fēci, -fectum, 3, *accomplish*
confīdo, -fīsus, 3, semi-dep. *have confidence*
confingo, -finxi, -fictum, 3, *invent*
confirmo, 1, *assure, confirm*
conflo, 1, *stir up*
cōnicio, -iēci, -iectum, 3, *throw to, infer, interpret, direct or apply to*
coniungo, -iunxi, -iunctum, 3, *join*
conlēga, -ae, m. *a companion*
conloco, 1, *set up, erect*
conlūceo, 2, *shine brightly*
cōnor, 1, dep. *attempt*
conquīro, -quīsīvi, -quīsītum, 3, *seek out*
conscrībo (scrībo), 3, *enrol*; **patrēs conscriptī**, *senators*
consecro, 1, *dedicate*
consequor, -sequi, secūtus, 3, dep. *overtake*
consīdero, 1, *consider*
consilium, -ii, n. *plan, intention, jury*
consisto, -stiti, -stitum, 3, *stop, am at rest, consist of, depend on*
conspectus, -ūs, m. *sight*
conspicio, -spexi, -spectum, 3, *look at*

constat, 1, impers. *it is evident, it is agreed*
consterno, -strāvi, -strātum, 3, *cover, pave*; **constrata** (navis), *having a deck*
constituo, -stitui, -stitūtum, 3, *set up, decide*
consuesco, -suēvi, -suētum, 3, *am accustomed*
consuētūdō, -inis, f. *custom*
consūmo, -sumpsi, -sumptum, 3, *spend, waste*
contendo, -tendi, -tentum, 3, *strive for*
contentus, adj. *content*
contestor, 1, dep. *call to witness*
continentia, -ae, f. *moderation*
contineo, -tinui, -tentum, 2, *contain*; (in passive) *am composed of, bounded by, consist of*
continuō, adv. *at once*
continuus, adj. *uninterrupted, one after another*
contrā, adv. and prep. *opposite, against*
contrōversia, -ae, f. *dispute*
conturbo, 1, *throw into confusion*
convenio (venio), 4, *come together, address, meet*
conventus, -ūs, m. *company, community, assembly, court of justice*
converto (verto), 3, *reverse*
convīcium, -ii, n. *noise, sound of wrangling*
convinco, 3, *convict, prove guilty*
convīvium, -ii, n. *a feast*
cooperio, -ui, -rtum, 4, *cover wholly, bury*
cōpiōsus, adj. *well supplied, eloquent*
cōram, adv. and prep. *publicly, in the presence of*
corōna, -ae, f. *wreath*
corrumpo, -rūpi, -ruptum, 3, *spoil, forge*
cotīdiē, adv. *daily*
crāpula, -ae, f. *intoxication*
crātēra, -ae, f. *mixing bowl*
crēbrō, adv. *frequently*
creo, 1, *create*
crepīdo, -inis, f. *pier, quay, sea-wall*
crīmen, -inis, n. *a charge*
crīminor, 1, dep. *accuse*
cruciātus, -ūs, m. *torture, execution*
crucio, 1, *torture*
crūdēlis, adj. *cruel*
crux, crucis, f. *cross*
cubiculum, -i, n. *a bedroom*
culpa, -ae, f. *blame, fault*
cum, conj. *when, since, although*
cunctus, adj. *all*
cupiditās, -ātis, f. *greed*
Cupīdo, -inis, m. *Cupid*
cupidus, adj. *desirous, selfish, greedy*
cupio, cupīvi, cupītum, 3, *wish for, desire strongly*
cūro, 1, *take care, see to*
curro, cucurri, cursum, 3, *run*
cursus, -ūs, m. *course*
custōs, -ōdis, c. *a guard*

damnātiō, -ōnis, f. *condemnation*
damno, 1, *condemn*
dē, prep. + abl. *from, concerning*
dēbeo, 2, *owe, ought*
dēcīdo, -cīdi, -cīsum, 3, *decide, settle*
dēcrētum, -i, n. *a decision, ordinance*
dēdūco (dūco), 3, *lead away*
dēfendo, -fendi, -fensum, 3, *defend*
dēfensiō, -ōnis, f. *defence*
dēfensor, -ōris, m. *a protector, supporter*

dēfero (fero), 3, *bring, offer, report*; **dēfero nōmen**, *denounce, accuse*
dēficio, -fēci, -fectum, 3, *break loose, desert, fail*
dēfīnio, -īvi, -ītum, 4, *set bounds to*
dēformo, 1, *spoil, dishonour*
deinde, adv. *then*
dēlecto, 1, *please*
dēligo, 1, *bind*
dēligo, dēlēgi, dēlectum, 3, *choose*
dēmentia, -ae, f. *madness*
dēmōlior, -ītus, 4, dep. *cast down, remove*
dēmonstrātiō, -ōnis, f. *a showing*
dēmonstro, 1, *prove, show*
dēnique, adv. *at last*
dēperdo, -didi, -ditum, 3, *lose, destroy*
dēporto, 1, *take away*
dēprecor, 1, dep. *beg*
dēprimo, -pressi, -pressum, 3, *sink, weigh down*
dēscendo, -ndi, -nsum, 3, *go down*
dēscrībo (scrībo), 3, *write down*
dēsīdero, 1, *long for, require*
dēsipio, 3, *am foolish*
dēspērātus, adj. *desperate*
dēspicio, -spexi, -spectum, 3, *look down upon*
dēstituo, -ui, -ūtum, 3, *leave alone*
dēsum, dēesse, dēfui, *am wanting*
dētraho, -traxi, -tractum, 3, *drag down, off*
deus, -i, m. *god*
diciō, -ōnis, f. *authority*
dico, 1, *dedicate, devote*
dīco, dixi, dictum, 3, *say, speak, plead*
difficultās, -ātis, f. *difficulty, embarrassment*
dignitās, -ātis, f. *dignity, position, character*
dīligens, adj. *hard-working*
dīligenter, adv. *carefully*
dīmissiō, -ōnis, f. *dismissal, sending away*
dīmitto (mitto), 3, *send away, let go, discharge*
dīreptiō, -ōnis, f. *pillaging, plundering*
dīripio, -ui, -eptum, 3, *tear in pieces, plunder*
discēdo, -cessi, -cessum, 3, *go away*
disciplīna, -ae, f. *teaching, discipline, rule*
disco, didici, 3, *learn*
dīscrībo (scribo), 3, *apportion, allot, dispense*
dispersē, adv. *here and there*
displiceo, -ui, -itum, 2, *displease, dissatisfy*
disputo, 1, *discuss, argue*
dissimulātiō, -ōnis, f. *pretence*
dissolvo, -solvi, -solūtum, 3, *refute, disprove*
distinctus, adj. *adorned*
distribuo, -ui, -ūtum, 3, *distribute*
diū, adv. *for a long time*
diūturnus, adj. *of long standing*
dīvārico, 1, *stretch apart*
doceo, 2, *teach*
doleo, dolui, dolitum, 2, *grieve*
dolor, -ōris, m. *grief*
domesticus, adj. *private*
dōnec, conj. *until*
dormio, 4, *sleep*
dubitātiō, -ōnis, f. *doubt*
dubito, 1, *doubt*
dubius, adj. *doubtful*
dūco, duxi, ductum, 3, *draw, lead, consider*
dux, ducis, c. *chief, leader*

ēbrius, adj. *drunk*
ebur, -oris, n. *ivory*

eburneus, adj. *of ivory*
ecce, interjection, *behold*
ēdīco (dīco), 3, *decree*
ēdūco (dūco), 3, *lead forth, summon to court*
effugio, -fūgi, 3, *escape*
egens, egentis, adj. *needy*
egestas, -ātis, f. *want*
ēgregiē, adv. *excellently*
ēicio, ēiēci, ēiectum, 3, *throw out, expel, banish*
ēlabōro, 1, *exert myself, endeavour*
ēlegans, adj. *tasteful*
ēligo, -lēgi, -lectum, 3, *choose*
ēluo, -ui, -ūtum, 3, *wash out*
ēmentior, ēmentītus, 4, dep. *pretend, utter falsely*
ēmigro, 1, *depart from*
emptiō, -ōnis, f. *purchase*
enim, conj. *for*
ēnumero, 1, *recount*
eo, ii or īvi, itum, *go*
eques, equitis, m. *a knight* (member of the Equestrian order)
equester, adj. *equestrian*
ēreptor, -ōris, m. *plunderer*
ergā, prep. + acc. *towards*
ērigo, -rexi, -rectum, 3, *lift up*
ēripio, -ripui, -reptum, 3, *snatch, tear away*
erro, 1, *wander, am mistaken*
ērumpo, -rūpi, -ruptum, 3, *break out*
ēsurio, -ītum, 4, *am hungry*
etenim, conj. *for indeed*
ēverto (verto), 3, *overturn*
ēvolo, 1, *fly away*
exaro, 1, *raise, produce by tillage*
excavo, 1, *hollow out*
excello, -cellui, -celsum, 3, *excel, surpass*
excipio, -cēpi, -ceptum, 3, *receive*
excito, 1, *awake*
excōgito, 1, *think out*
exemplum, -i, n. *example, precedent*
exeo (eo), *leave, go out*
exhibeo, 2, *present, show*
exigo, -ēgi, -actum, 3, *demand, exact*
eximius, adj. *select, extraordinary*
eximo, -ēmi, -emptum, 3, *release, remove*
exinānio, -īvi, -ītum, 4, *empty*
existimātiō, -ōnis, f. *opinion, reputation*
existimo, 1, *think, reckon*
exitus, -ūs, m. *end*
exorno, 1, *embellish, fit out*
expedio, 4, *set free*; expedit (impers.), *is useful, advantageous*
expīlo, 1, *rob*
expio, 1, *purify, make amends*
expōno (pōno), 3, *reveal, explain*
expressus, adj. *clear*
exsilio, -ui, 4, *spring out*
exspectātiō, -ōnis, f. *awaiting, speculation*
exspecto, 1, *wait for*
exsul, -ulis, c. *exile*
extenuo, 1, *diminish, weaken, palliate*
exteri, plur. adj. *foreign*
extorqueo, -si, -tum, 2, *wrest away*
exturbo, 1, *drive out*

facile, adv. *easily*
facinus, -oris, n. *act, outrage, crime*
facio, fēci, factum, 3, *make, do*
facultās, -ātis, f. *capability, opportunity*
fallo, fefelli, falsum, 3, *disappoint, deceive*
fāma, -ae, f. *report, public opinion, fame*

famēs, -is, f. *hunger*
familia, -ae, f. *family, household*
familiāritās, -ātis, f. *intimacy*
fānum, -i, n. *temple*
farcio, farsi, fartum, 4, stuff
fascis, -is, m. *bundle*; (in plur.) *the rods and axe carried before highest magistrates*
Favōnius, -ii, m. *the West wind*
fēlīciter, adv. *happily, successfully*
fera, -ae, f. *a wild beast*
ferē, adv. *almost*
ferio, 4, *strike*; + **secūrī** = *execute*
fero, ferre, tuli, lātum, irr. *carry, bear, endure*
fero molestē, *am vexed*
ferrum, -i, n. *sword*
festīnātiō, -ōnis, f. *haste*
festus, adj. *joyful*; **diēs festi**, *holidays*
fidēlis, adj. *faithful*
fidēs, -ēi, f. *faith, honour, loyalty*
figūra, -ae, f. *shape*
fīlius, -ii, m. *son*
fingo, finxi, fictum, 3, *make, invent*
fīnis, -is, m. *end*
fīo, fieri, factus, 3, semi-dep. *become, happen, am done*
flāgitium, -ii, n. *disgraceful behaviour*
fleo, -ēvi, -ētum, 2, *weep*
fluctuo, 1, *am driven hither and thither*
fluctus, -ūs, m. *wave*
fore, future infinitive of sum
formīdolōsus, adj. *dreadful*
formōsus, adj. *shapely*
fortasse, adv. *perhaps*
forte, adv. *perchance*
fortūna, -ae, f. *fate, fortune*
frango, **frēgi**, fractum, 3, *break*
frequens, adj. *numerous, in large numbers*
frequentia, -ae, f. *crowd, great number*
frīgus, -oris, n. *cold*
frūges, -um, f. *corn*
frūmentārius, adj. *of corn*; **rēs frūmentāria**, *corn supply*
frūmentum, -i, n. *corn*
fuga, -ae, f. *flight*
fugitīvus, -i, m. *runaway slave*
fulgor, -ōris, m. *brightness*
fūmo, 1, *smoke*
fundo, fūdi, fūsum, 3, *pour out, prostrate*
fūnestus, adj. *deadly, fatal*
fūr, fūris, m. *thief*
furo, 3, *am mad*
furor, -ōris, m. *madness, frenzy*
fūror, 1, dep. *steal*
furtum, -i, n. *theft*

gaudeo, gāvīsus, 2, dep. *rejoice*
gemitus, -ūs, m. *groan*
gemma, -ae, f. *a gem, precious stone*
gens, -tis, f. *race, family*
genus, -eris, n. *kind, species, race*
gero, gessi, gestum, 3, *do, carry out*; **sē gerere**, *to behave*
Graeculus, adj. and noun, *a mere Greek*
grandis, adj. *large, great*
grātia, -ae, f. *love, friendship, influence*
grātus, adj. *pleasant, thankful*
gravis, adj. *heavy*

habeo, 2, *have, consider*
habito, 1, *live in, inhabit*
habitus, -ūs, m. *appearance*
haruspex, -icis, m. *soothsayer*
hērēditās, -ātis, f. *inheritance*
hērēs, -ēdis, c. *master or owner*

hibernus, adj. *winter*
hicce, haecce, hōcce, emphatic form of hic, haec, hōc
hiems, -mis, f. *winter*
hodiē, adv. *to-day*
honestus, adj. *honourable, distinguished*
honōrificē, adv. *with honour*
hōra, f. *hour, time, season*
hortor, 1, dep. *incite*
hospes, -itis, m. *host, guest, friend*
hospitium, -ii, n. *hospitality*
hostīlis, adj. *like or belonging to an enemy*
hūmānitās, -ātis, f. *kindness, refinement, noble feelings*
hydria, -ae, f. *water-jug*

iaceo, -cui, 2, *lie*
iacto, 1, *throw*; **sē iactāre**, *to boast*
iam, adv. *now, already*
iamdūdum, adv. *a long time ago*
iānitor, -ōris, m. *doorkeeper*
ictus, ictūs, m. *blow*
idcircō, adv. *on that account*
idiōta, -ae, m. *uneducated or inexperienced person*
idōneus, adj. *fitting, suitable*
ignis, -is, m. *fire*
ignōminia, -ae, f. *disgrace*
ignōro, 1, *am unaware of*
ignōtus, adj. *unknown*
illūc, adv. *to that place*
illustris, adj. *famous, distinguished*
imber, -bris, m. *shower of rain, rain*
immānis, adj. *monstrous, immense*
imminuo, 3, *diminish*
impedio, 4, *hinder*
imperātor, -ōris, m. *commander*
imperium, -ii, n. *authority, supreme power*
impero, 1, *command, rule over*
impertio, -īvi, -ītum, 4, *tell to another*
impetro, 1, *obtain a request, get leave, effect*
impetus, -ūs, m. *attack*
impiē, adv. *wickedly, profanely*
impiger, adj. *active, energetic*
implōro, 1, *beg for, implore*
importūnitās, -ātis, f. *rudeness, insolence, wickedness*
improbitās, -ātis, f. *wickedness, dishonesty*
improbo, 1, *reject, make void*
improbus, adj. *wicked, dishonest, shameless*
imprōvīsus, adj. *unforeseen*
imprūdentia, -ae, f. *recklessness, inconsiderateness*
impudentia, -ae, f. *impudence, effrontery*
impūne, adv. *safely, without punishment*
impūnītus, adj. *unpunished*
impūrus, adj. *filthy, nasty*
inambulo, 1, *walk up and down*
inānis, adj. *empty*
inaudītus, adj. *unheard of*
incendo, -di, -sum, 3, *kindle, irritate*
inceptum, -i, n. *an undertaking*
incipio, -cēpi, -ceptum, 3, *begin*
inclūdo (claudo), 3, *shut up, hide*
incognitus, adj. *untried*
incolo, -ui, 3, *inhabit*
incommodum, -i, n. *trouble, inconvenience*
incrēdibilis, adj. *unbelievable*
incurvus, adj. *bent*
indicium, -ii, n. *sign, proof*
indico, 1, *point out, show*
indīco, -dixi, -dictum, 3, *declare publicly, proclaim*; + **bellum**, *declare war*; **indictā causā**, without a hearing

indignus, adj. *unworthy*
induo, -ui, -ūtum, 3, *clothe, entangle*
iners, adj. *indolent*
infāmia, -ae, f. *disgrace*
infero (fero), 3, *bring forward*; **crīmen infero**, *make a charge against*
inflammo, 1, *kindle, arouse, set on fire*
ingeniōsus, adj. *intelligent, having good natural abilities*
ingenium, -ii, *nature, inclination, capability*
inicio, -iēci, -iectum, 3, *throw upon*
inimīcitia, -ae, f. *hostility*
inimīcus, adj. and noun, *hostile, enemy*
inīquus, adj. *unjust, unfavourable*
initium, -ii, n. *beginning*
iniūria, -ae, f. *wrong, injustice, insult*
iniussū (only in abl.), *without the order of*
inlustro, 1, *illuminate*
innocens, adj. *harmless, blameless*
innumerābilis, adj. *very many, countless*
inopīnātus, adj. *unexpected*
inops, adj. *helpless*
inquam, **inquit**, defective verb, *I say, says he*
inquīro, -sīvi, -sītum, 3, *seek for* (evidence)
insānia, -ae, f. *madness, extravagance*
insignis, adj. *remarkable, especial, illustrious*
instar, indecl. noun, n. *appearance*
instinguo, -stinxi, -stinctum, 3, *impel*
instituo, -ui, -ūtum, 3, *set up, establish, begin*
institūtum, -i, n. *plan, practice, arrangement*
instruo, -struxi, -structum, 3, *draw up, furnish*
insula, -ae, f. *island*
integer, adj. *whole, honest, undecided*
intellegens, adj. *understanding*
intellego, -xi, -ctum, 3, *understand*
intempestus, adj. *unseasonable*; **nocte intempestā**, *at dead of night*
intendo, -di, -tum and -sum, *stretch out*
intereā, adv. *meanwhile*
interlino, -lēvi, -litum, 3, *falsify by erasing*
interpello, 1, *interrupt, disturb*
interpres, -etis, m. *agent, negotiator*
interrogo, 1, *question, examine*
intolerandus, adj. *unbearable*
intrō, adv. *within*
introitus, -ūs, m. *entrance*
intueor, 2, dep. *look at*
inveho, -vexi, -vectum, 3, *carry into a place*; (in passive) *attack with words*
invenio (venio), 4, *find, discover*
invesperascit, 3, impers. *evening approaches*
invidia, -ae, f. *ill-will, unpopularity*
invīto, 1, *invite*
invītus, adj. *unwilling*
involūcrum, -ii, n. *a wrapping*
involvo, -volvi, -volūtum, 3, *wrap up*
īrācundia, -ae, f. *rage*
īrascor, **īrātus**, 3, dep. *am angry*
iste, pron. *that of yours, your client*
itaque, conj. *and so*

ita, adv. *so, thus, on this condition*
item, adv. *likewise, also*
iterum, adv. *again*
iubeo, iussi, iussum, 2, *order*
iūcundus, adj. *pleasing*
iūdex, -icis, c. *judge, juryman*
iūdicium, -ii, n. *trial, judgment, legal verdict*
iūro, 1, *swear*; **iūrātus**, *a person under oath*
iūs, iūris, n. *right, law*
iustus, adj. *right, proper*
iuvo, iūvi, iūtum, 1, *help*; (impers.) **iuvat mē**, *I am delighted*

kalendae, -ārum, f. *first day of the month*

labōro, 1, *work, am anxious*
lacus, -ūs, m. *lake*
laetor, 1, dep. *rejoice*
laqueus, -i, m. *noose, chain*
Latīnē, adv. *plainly*
lātrōcinium, -ii, n. *robbery*
latus, -eris, n. *side, lungs*
lectīca, -ae, f. *litter*
lectus, -i, m. *bed*
lēgātiō, -ōnis, f. *lieutenancy, deputy governorship*
lego, lēgi, lectum, 3, *read*
lēniter, adv. *gently*
levis, adj. *light, worthless, unimportant*
lex, lēgis, f. *law*
libentissimē, adv. *very willingly*
Līber, -i, m. *Bacchus*
līberālis, adj. *gentlemanly, kindly*
līberi, -ōrum, m. *children*
lībero, 1, *set free, acquit*
lībertās, -ātis, f. *freedom, liberty*
lībertus, -i, m. *a slave who has been set free, a freedman*
libet, 2, impers. *it pleases*
libīdinōsus, adj. *licentious, self-willed*
libīdō, -inis, f. *wilfulness, caprice, lust*
licet, licuit, 2, impers. *it is allowed*
lictor, -ōris, m. *magistrate's attendant, lictor*
ligneus, adj. *of wood*
līnum, -i, n. *linen*
liquidō, adv. *plainly*
litterae, -ārum, f. *document, letter, records*
litūra, -ae, f. *erasure*
lītus, -oris, n. *sea-shore*
locuplēs, ētis, adj. *rich*
locuplēto, 1, *enrich*
locus, -i, m. *place, topic*
longitūdō, -inis, f. *length*
lūceo, luxi, 2, *shine*; lūcet, *it is light*
lucror, 1, dep. *gain, pocket*
luctuōsus, adj. *causing sorrow*
lūcus, -i, m. *grove*
lūdibrium, -ii, n. *mockery*
luo, lui, 3, *set free, atone for*
luteus, adj. *worthless*
luxuriēs, -ēi, or **luxuria**, -ae, f. *luxury, excess*

macula, -ae, f. *spot, hole*
maeror, -ōris, m. *sorrow*
magis, adv. *more*; **eō magis**, *all the more*
māiestās, -ātis, f. *dignity, high treason*
māior, comp. adj. *greater*; **māiōres**, *ancestors*
maleficium, -ii, n. *evil deed*
mālo, malle, mālui, *prefer*
mālus, -i, m. *a mast*
mancipium, -ii, n. *slave*
mando, 1, *order, entrust*
māne, adv. *early*
manifestus, adj. *obvious, clear*

manubrium, -ii, n. *handle*
margarīta, -ae, f. *pearl*
marmor, -oris, n. *marble*
marmoreus, adj. *of marble*
maximē, adv. (superlative) *very greatly*
medicus, -i, m. *doctor*
mediocriter, adv. *ordinarily*
meditor, 1, dep. *think upon, design*
medius, adj. *middle*
Melitensis, adj. *Maltese*
meminī, 3, defective verb, *remember*
memoria, -ae, f. *memory*
mendum, -i, n. *fault, error*
mens, mentis, f. *mind, wits*
mentiō, -ōnis, f. *mention*
mercātor, -ōris, m. *merchant*
mercēs, -ēdis, f. *pay, income*
metuo, -ui, -ūtum, 3, *fear*
metus, -ūs, m. *fear*
minae, -ārum, f. *threats*
minor, -ātus, 1, dep. *threaten*
minus, adv. *less*
minūtus, adj. *small*
mīrē, adv. *strangely, wonderfully, exceedingly*
mīror, 1, dep. *wonder, am surprised*
mīrus, adj. *astonishing*
misericordia, -ae, f. *pity*
missiō, -ōnis, f. *discharge*
mitto, mīsi, missum, 3, *send*
modo, adv. *only*
modus, -i, m. *way, kind, method*
moenia, -ium, n. *ramparts, walls*
molestē, adv. *with difficulty*; molestē **fero**, *am vexed*
mōlior, -ītus, 4, dep. *struggle*
mora, -ae, f. *delay*
morbus, -i, m. *disease, illness*
morior, mortuus, 3, dep. *die*
mors, mortis, f. *death*
mortālis, adj. and noun, *mortal*
mōs, mōris, m. *custom*
moveo, mōvi, mōtum, 2, *move*
muliebris, adj. *feminine, female*
mulier, -is, f. *woman*
multitūdō, -inis, f. *crowd*
multō, adv. *much*
mūnio, 4, *fortify*
mūnus, -eris, n. *gift*
myoparō, -ōnis, m. *small pirate ship*
mystagōgus, -i, m. *guide to sacred places*

nam, conj. *for*
nāris, -is, f. *nostril*
narro, 1, *tell, relate*
nascor, nātus, 3, dep. *am born*
nātiō, -ōnis, f. *tribe, people*
nauarchus, -i, m. *shipmaster*
nauta, -ae, m. *sailor*
nāvālis, adj. *connected with ships*
nāvigium, -ii, n. *boat*
nāvigo, 1, *sail*
nē . . . quidem, *not even*
necessāriō, adv. *necessarily, inevitably*
necesse, n. adj. *necessary*
neco, 1, *kill*
nefārius, adj. *wicked*
negōtior, 1, dep. *engage in, carry on business*
negōtium, -ii, n. *an affair, task*
nēmō, -inis, c. *no one*
nēquāquam, adv. *by no means*
neque, conj. *neither, and not*
nēquissimus, superlative of nēquam, *worthless*
nēquitia, -ae, f. *villainy*
nescio, -ii, -ītum, 4, *do not know*
nex, necis, f. *death*
niger, adj. *black, dark*
nihil, n. indecl. *nothing*
nīmīrum, adv. *no doubt*
nimis, adv. *too much*
nisi, conj. *unless*

nitor, **nisus**, and nixus, **3**, dep. *strive, rely on*
nōbilis, adj. *well known, well born*
nōbilitās, -ātis, f. *nobility, excellence*
nocens, adj. *guilty*
noceo, 2, *harm, injure*
nōlo, nolle, nōlui, *am unwilling*
nōmen, -inis, n. *name, fame*
nōndum, adv. *not yet*
nōnus, adj. *ninth*
nosco, nōvi, nōtum, 3, *know*
noto, 1, *observe, mark*
nūdo, 1, *strip, lay bare*
nūdus, adj. *naked, empty, stripped*
nūgātōrius, adj. *worthless*
nullus, adj. *none*
nūmen, -inis, n. *divine sway, deity*
numero, 1, *count, pay down*
nummārius, adj. *financial, of money*
nummus, -i, m. *coin, cash*
nunc, adv. *now*
nuntius, -ii, m. *messenger, message*
nūper, adv. *lately*
nusquam, adv. *nowhere*
nūtus, -ūs, m. *nod*

ob, prep. + acc. *on account of*
obeo (eo), *survey*
obicio, -iēci, -iectum, 3, *throw to, expose*
obrigesco, -rigui, 3, *become stiff*
obscūrus, adj. *unintelligible, doubtful*
obsecro, 1, *entreat*
obstringo, -nxi, -ctum, 3, *tie, bind*
obtempero, 1, *obey*
obtero, -trīvi, -trītum, 3, *crush*
obtestor, 1, *call to witness*
obtineo, -tinui, -tentum, 2, *possess, get possession of, assert*
obvenio (venio), 4, *fall in with, fall to the lot of*
obvolvo, -volvi, -volūtum, 3, *wrap up*
occīdo, -di, -sum, 3, *kill*
occultē, adv. *secretly*
occulto, 1, *hide*
octaphoron, -i, n. *a litter carried by eight bearers*
oculus, -i, m. *eye*
odium, -ii, n. *hatred, disgust*
odōror, 1, *smell out*
offendo, -di, -sum, 3, *shock, meet with, find*
officium, -ii, n. *a kindness, duty*
ōlim, adv. *once, formerly*
ōmen, -inis, n. *an omen*
omitto (mitto), 3, *leave out*
omnīnō, adv. *altogether, indeed*
onus, -eris, n. *burden*
opera, -ae, f. *labour, care*; **operam dare**, *to take care, see to it*
opēs, -um, plur. f. *wealth, power*
opīniō, -ōnis, f. *supposition, idea*
opīnor, 1, dep. *conjecture, consider*
oportet, -uit, 2, impers. *it is necessary, proper, reasonable*
oppōno, -posui, -positum, 3, *oppose, set against*
opportūnus, adj. *fit, serviceable*
oppugno, 1, *attack*
opto, 1, *choose, desire*
opulens, adj. *rich*
opus, -eris, n. *work*; **opus est**, *there is need*
ōrātiō, -ōnis, f. *speech*
orbis, -is, m. *circle*; **orbis terrārum**, *the world*

ordō, -inis, m. *class, order, rank*
ornāmentum, -i, n. *decoration*
orno, 1, *equip*
ōro, 1, *beg*
ōs, **ōris**, n. *mouth, face, impudence*
ostendo, ostendi, ostentum and -sum, 3, *show*
ostium, -ii, n. *small entrance, mouth of a harbour*

paene, adv. *almost*
palaestra, -ae, f. *place of exercise, wrestling-school*
palam, adv. *openly*
pallium, -ii, n. *cloak*
palma, -ae, f. *palm*
pālus, -i, m. *stake*
pār, paris, adj. *equal*
parco, peperci, parsum, 3, *spare*
parens, -ntis, m. *parent*
pāreo, 2, *obey*
pariēs, -etis, m. *room-wall*
paro, 1, *prepare*
partim, adv. *partly*
parvulus, adj. *very small*
patefacio (facio), 3, *lay open, expose*
pateo, -ui, 2, *am open, available*
patior, passus, 3, dep. *suffer*
patrius, adj. *of a father, paternal*
patrōnus, -i, m. *protector, defender, advocate*
paulisper, adv. *for a little, for a short time*
paulō, adv. *a little, somewhat*
paulum, adv. *a little*
pecco, 1, *do wrong, make a mistake*
pecuārius, -i, m. *a grazier, stock farmer*
pecūnia, -ae, f. *money*
pendo, pependi, pensum, 3, *weigh, consider*
penes, prep.+acc. *in possession of*
penetro, 1, *enter*
pēnūria, -ae, f. *lack, scarcity*
per, prep.+acc. *through, by means of*
peracūtus, adj. *very clever*
perantīquus, adj. *very old*
percrepo, -ui, 1, *ring, resound*
percutio, -cussi, -cussum, 3, *strike*
perditus, adj. *wicked, abandoned, hopeless, desperate*
perdo, -didi, -ditum, 3, *lose, destroy*
perdūco (dūco), 3, *induce*
perfectus, adj. *exquisite*
perficio (facio), 3, *carry out, accomplish*
perfidia, -ae, f. *dishonesty, treachery*
perfrīgidus, adj. *very cold*
perfugium, -ii, n. *refuge, resort*
pergrandis, adj. *very large*
perīculum, -i, n. *danger*
perītus, adj. *skilful, experienced*
perlūcidus, adj. *transparent*
perōro, 1, *bring a speech to an end, finish* (speaking)
perpetuō, adv. *for ever*
perpōto, 1, *keep drinking*
perrīdiculē, adv. *very laughably*
perscrūtor, 1, dep. *search thoroughly*
persequor, -secūtus, 3, dep. *pursue, avenge*
persevēro, 1, *proceed with steadily, persist*
persōna, -ae, f. *a mask, rôle*
perspicio, -spexi, -spectum, 3, *observe, gaze at*
perspicuus, adj. *clear, obvious*
pertineo, -tinui, -tentum, 2, *concern*
perturbo, 1, *confuse, disturb*

pervagor, 1, dep. *wander through*; **pervagātus**, *well known*
pervenio (venio), 4, *arrive*
pervestīgo, 1, *search out*
pervetus, adj. *very old*
pervolo, -velle, -volui, *wish greatly*
pervulgo, 1, *publish, spread abroad*
peto, petīvi, petītum, 3, *seek*
pictor, -ōris, m. *painter*
pietās, -ātis, f. *sense of duty, affection, reverence*
pingo, pinxi, pictum, 3, *paint*
pīrāta, -ae, m. *pirate*
plācābilis, adj. *easily appeased*
placeo, 2, *please*
placidē, adv. *calmly*
plānē, adv. *openly, clearly, evidently*
plērīque, plur. adj. *most*
pōculum, -i, n. *cup*
poena, -ae, f. *penalty, punishment*
Poena, -ae, f. *avenging deity*
polliceor, pollicitus, 2, dep. *promise*
pondus, -eris, n. *weight, importance*
pōno, posui, positum, 3, *set up, erect, place, state*
populus, -i, m. *people*
porrō, adv. *furthermore*
porticus, -ūs, f. *porch*
portus, -ūs, m. *harbour*
posco, poposci, 3, *demand*
possum, posse, potui, irr. *am able, can, am powerful*
posteā, adv. *afterwards*
posterius, adv. *later*
posthāc, adv. *after this*
postrēmus, adj. *last*
postrīdiē, adv. *on the next day*
postulo, 1, *demand*
potens, -entis, adj. *powerful*
potestās, -ātis, f. *power, authority*
potissimum, adv. *especially*
potius, adv. *rather*
prae, prep. + abl. *before, in comparison with, because of*
praebeo, 2, offer, show; **sē praebēre**, *to behave oneself*
praeceps, -cipitis, adj. *headlong*
praecīdo, -cīdi, -cīsum, 3, *cut through*
praecipuē, adv. *specially*
praeclārē, adv. *very plainly, excellently*
praeclārus, adj. *glorious, distinguished*
praecō, -ōnis, m. *herald*
praeda, -ae, f. *booty, spoil*
praeditus, adj. *gifted, endowed*
praedō, -ōnis, m. *robber, plunderer*
praedor, 1, dep. *plunder*
praeficio, -fēci, -fectum, 3, *place in authority*
praepōno (pōno), 3, *place in command of*
praesentiā (**in**), adv. phrase *at the moment*
praesideo, -sēdi, 2, *protect*
praestō, adv. *at hand*
praesum, -esse, -fui, *am set over*
praeter, prep. + acc. *beyond, except*
praetereā, adv. *besides, as well*
praetereo, -ii, -itum, *leave out*
praetermitto (mitto), 3, *pass over, leave out, omit*
praetervehor, -vectus, 3, dep. *sail by*
praetor, -ōris, m. *praetor*
praetōrium, -ii, n. *governor's house*
praetūra, -ae, f. *praetorship*
precem, precī, prece, f. defective, *prayer*
pretium, -ii, n. *price, bribe*; **operae pretium est**, *it is worth while*

prīmō, adv. *at first*
prīmum, adv. *in the first place*
princeps, -ipis, c. *first, chief, leader*
principium, -ii, n. *beginning*
prius, adv. *before*
prīvātus, adj. *private*
prīvo, 1, *deprive of, strip*
prō, exclamation, *Oh!*
prō, prep. + abl. *in front of, instead of, for the sake of, in view of*
proāgorus, -i, m. *director* (chief magistrate in certain Sicilian towns)
probo, 1, *approve, demonstrate, justify*
prōcrastino, 1, *hesitate, delay*
prōditiō, -ōnis, f. *betrayal, treachery*
prōdo, -didi, -ditum, 3, *report, betray*
profānus, adj. *secular, not sacred*
profectō, adv. *certainly*
prōfero (fero), *display, adjourn*
prōficio, -fēci, -fectum, 3, *make progress*
proficiscor, profectus, 3, dep. *set out*
profiteor, professus, 2, dep. *promise*
prōfugio, -fūgi, 3, flee
prohibeo, 2, *forbid, prevent*
proinde, adv. *and so, wherefore*
prōmo, prompsi, promptum, 3, *bring out*
prōnuntio, 1, *publish, pronounce a judgment*
prope, adv. *nearly, almost*
prope, prep. + acc. *near*
propero, 1, *hurry*
propinquus, -i, m. *relative, kinsman*
prōpōno (pōno), 3, *set before*
propter, prep. + acc. *on account of*
prōpugnāculum, -i, n. *defence*
prōpugnātor, -ōris, m. *defender, marine*
prorsus, adv. *certainly, utterly*
prōsequor, -secūtus, 3, dep. *follow, pursue*
prout, adv. *just as, in proportion as*
prōveho (veho), 3, *carry forwards*; (in passive), *sail*
prōvideo (video), 2, *foresee, pay attention*
proximus, adj. *nearest, very near*
publicus, adj. *belonging to the state, public, common*
pudor, -ōris, m. *decency, good manners*
puerīlis, adj. *childish*
pulchritūdō, -inis, f. *beauty*
pulvīnus, -i, m. *cushion*
pūrus, adj. *clean, free from*
puto, 1, *think, consider*

quadrirēmis, -is, f. *vessel with four banks of oars*
quaero, quaesīvi, quaesītum, 3, *seek, ask for, investigate*
quaeso, 3, *beg*
quaestiō, -ōnis, f. *enquiry, court*
quaestūra, -ae, f. *quaestorship*
quaestus, -ūs, m. *gain*
quam prīmum, adv. *as soon as possible*
quāpropter, adv. *wherefore*
quārē, adv. *why?*
quasi, conj. *as if*
querimōnia, -ae, f. *complaint*
queror, questus, 3, dep. *complain*
quid, used as adv. *why?*
quīdam, pron. and adj. *a certain* (person)
quidem adv. *indeed*

quiesco, quiēvi, quiētum, 3, *keep quiet, sleep*
quīn, conj. + subjunctive, *but that, that*
quisnam, interr. pron. *who, pray?*
quispiam, pron. *anyone*
quisquam, pron. *anyone*
quisque, pron. and adj. *each*
quisquis, pron. and adj. *whoever, whatever*
quīvīs, pron. and adj. *anyone, any whatever*
quō, adv. *whither, to which place*
quod, conj. *because*
quodsī, conj. *but if*
quōnam, adv. *whither, pray?*
quoniam, conj. *since*
quoque, adv. *also*

rādix, -īcis, f. *root*
rapīna, -ae, f. *robbery*
rapio, -ui, -tum, 3, *snatch, seize, carry off*
ratiō, -ōnis, f. *consideration, principle, reason, manner, method*
recens, adj. *new, recent*
receptāculum, -i, n. *a place of refuge, harbour*
recipio, -cēpi, -ceptum, 3, *receive, undertake, take back*
reclāmo, 1, *cry out against, protest*
recognosco, -nōvi, -nitum, 3, *recollect*
recreo, 1, *restore*
recūso, 1, *refuse, deny*
reddo, -didi, -ditum, 3, *give back, hand over*
redeo (eo), *come back*
redimo, -ēmi, -emptum, 3, *buy back, purchase*
redundo, 1, *overflow*
refero, rettuli, relātum, 3, *carry back, refer, record*
refertus, adj. *stuffed*
refugio, -fūgi, 3, *escape*
reicio, reiēci, reiectum, 3, *throw back, refer*
religiō, -ōnis, f. *worship, rites, religion, reverence*
religiōsus, adj. *devout, scrupulous*
relinquo, relīqui, relictum, 3, *leave*
reliquus, adj. *remaining*
remedium, -ii, n. *remedy*
rēmex, -igis, m. *oarsman*
removeo (moveo), 2, *remove, separate*
rēmus, -i, m. *oar*
renuntio (nuntio), 1, *refuse, break off*
reor, ratus, 2, dep. *think*; **ratus**, *valid*
repello, reppuli, repulsum, 3, *ward off*
repente, adv. *suddenly*
reperio, repperi, repertum, 4, *find*
repōno (pōno), 3, *replace, place, set*
reporto, 1, *take back*
repugno, 1, *resist*
requīro, -quīsīvi, -quīsītum, 3, *seek for, need*
rescindo, -scidi, -scissum, 3, *repeal*
reservo (servo), 1, *save, keep back*
resisto, -stiti, 3, *oppose*
respergo, -si, -sum, 3, *sprinkle*
respondeo, respondi, responsum, 2, *answer*
restituo, -stitui, -stitūtum, 3, *restore, give back*
rēticulum, -i, n. *net-work bag*
retineo, -tinui, -tentum, 2, *retain*
retrūsus, adj. *concealed*
reus, -i, m. *an accused person, defendant*
revello, -velli, -vulsum, 3, *tear off*

reverto and **revertor** (verto), 3, *come back*
revoco, 1, *call back, restrain*
rogo, 1, *ask*
rudis, adj. *raw, inexperienced, ignorant*

sacer, adj. *holy*
sacrārium, -ii, n. *shrine, private chapel*
sacrilegus, -i, m. *temple-robber; one who commits sacrilege*
sacro, 1, *sanctify*
sacrum, -i, n. *sacred place or thing*; (in plur.), *sacred rites*
saeculum, -i, n. *age, generation*
saepe, adv. *often*; **saepenumerō**, *oftentimes*
sagum, -i, n. *blanket, military cloak*
salum, -i, n. *sea*
salūs, -ūtis, f. *safety, salvation, acquittal*
sānē, adv. *naturally, certainly, of course*
sanguis, -inis, m. *blood*
sāno, 1, *heal, correct, repair*
sapio, -īvi, 3, *am wise*
satietās, -ātis, f. *satiety*
satio, 1, *satisfy*
satis, adv. *enough, satisfactorily*
scelus, -eris, n. *crime, wickedness*
scio, 4, *know*
scītē, adv. *skilfully*
scrība, -ae, m. *clerk*
scrībo, scripsi, scriptum, 3, *write*
scyphus, -i, m. *cup*
sēcrētō, adv. *in private*
secūris, -is, f. *axe*; **percutio secūrī**, *behead*
sedeo, sēdi, sessum, 2, *sit*
sella, -ae, f. *magistrate's seat*
semel, adv. *once*
senātūs consultum, -i, n. *decree of the senate*
senīlis, adj. *old*
sensus, -ūs, m. *feeling, sensation*
sententia, -ae, f. *opinion, vote, verdict*
sentio, sensi, sensum, 4, *feel, perceive*
sepelio, -īvi, -ultum, 4, *bury*
sepultūra, -ae, f. *burial*
sequester, -tri, m. *agent, go-between*
sequor, secūtus, 3, dep. *follow, pursue*
sermō, -ōnis, m. *talk, gossip*
servīlis, adj. *like a slave*
servo, 1, *preserve, keep*
sevērus, adj. *strict, austere*
sīcārius, -ii, m. *assassin*
Siciliensis, adj. *Sicilian*
Siculus, adj. and noun, *of Sicily, a Sicilian*
sīcut, conj. *just as*
sigillātus, adj. *adorned with little images*
significo, 1, *show, exhibit*
signum, -i, n. *signal, statue, picture*
silentium, -ii, n. *quiet*
simul, adv. *at the same time*
simulācrum, -i, n. *image, portrait, statue*
simulātiō, -ōnis, f. *pretence*
simul atque, conj. *as soon as*
singulāris, adj. *unique, unparalleled, matchless*
sino, sīvi, situm, 3, *allow*
situs, -ūs, m. *position*
sīve . . . sīve, conj. *whether . . . or*
societās, -ātis, f. *fellowship, alliance*
socius, -ii, m. *ally*
sodālis, -is, c. *crony*
soleātus, adj. *wearing sandals*
soleo, solitus, 2, semi-dep. *am accustomed to*

sōlum, adv. *only*
sōlus, adj. *alone, only*
somnus, -i, m. *sleep*
sors, sortis, f. *lot*
speciēs, -ēi, f. *outward show*
spectāculum, -i, n. *sight*
specto, 1, *look at, incline to*
specula, -ae, f. *watch-tower*
spēro, 1, *hope*
splendidus, adj. *distinguished*
splendor, -ōris, m. *brilliance*
spoliātiō, -ōnis, f. *pillaging, plundering*
spolio, 1, *despoil, spoil*
spolium, -ii, n. *spoil*
statim, adv. *at once*
statua, -ae, f. *statue*
statuo, -ui, -ūtum, 3, *determine, decide*
status, -ūs, m. *condition*
stilus, -i, m. *pen*
stīpendium, -ii, n. *pay, salary*
stīpo, 1, *surround*
stirps, -pis, f. *root*
strepitus, -ūs, m. *noise, din*
studium, -ii, n. *zeal, enthusiasm*
subcrispus, adj. *rather curly*
subicio, subiēci, subiectum, 3, *prompt*
subitō, adv. *suddenly*
subripio, -ripui, -reptum, 3, *snatch away secretly*
subsellium, -ii, n. *bench, court*
subsidium, -ii, n. *support, reserve*
substituo, -ui, -ūtum, 3, *put in place of*
subterfugio, -fūgi, 3, *escape*
succēdo, -cessi, -cessum, 3, *succeed, follow*
summa, -ae, f. *main thing, chief point, sum of money*
summus, adj. *highest, very high*
sūmo, sumpsi, sumptum, 3, *take, choose*
sumptus, -ūs, m. *expense*
supellex, -ectilis, f. *furniture*
superbia, -ae, f. *pride, arrogance*
supersum, -esse, -fui, *am left, remain over*
superus, adj. *upper, higher*
suppedito, 1, *supply*
supplicium, -ii, n. *punishment, torture*; **supplicium sūmo**, *inflict punishment*
suppōno (pōno), 3, *substitute*
suscipio, -cēpi, -ceptum, 3, *undertake*
suspīciō, -ōnis, f. *suspicion*
suspicor, 1, dep. *suspect*
sustineo, -tinui, -tentum, 2, *sustain, hold up*
symphōnia, -ae, f. *harmony, music*
symphōniacus, adj. *musical*; symphoniaci homines, *bandsmen, choristers*
Syrācūsae, -ārum, f. *Syracuse*

tabernāculum, -i, n. *tent*
tabula, -ae, f. *picture*; (in plur.) *records, papers*
taceo, -ui, -itum, 2, *am silent*
tacitē, adv. *silently*
tālāris, adj. *reaching to the ankles*
tam, adv. *so*
tamen, adv. *but, however, yet*
tametsī, conj. *although*
tamquam, conj. *as if*
tantulum, -i, n. *so little*
tantum, adv. *only*
tectum, -i, n. *roof, house*
tego, -xi, -ctum, 3, *cover*
tēlum, -i, n. *weapon*
tempero, 1, *show moderation to, consideration for*
tempestās, -ātis, f. *weather, storm, calamity*
templum, -i, n. *temple*
tempus, -oris, n. *time, occasion, season*

tenebrae, -ārum, f. *darkness, obscurity*
teneo, -ui, -tum, 2, *hold, know*
tenuis, adj. *thin*
tergum, -i, n. *back*
terra, -ae, f. *earth, ground*
tertius, adj. *third*
testimōnium, -ii, n. *evidence*
testis, -is, c. *witness*
testor, 1, dep. *declare, bear witness to*; **testātus**, *public, manifest*
textīle, -is, n. *fabric, tapestry*
tollo, sustuli, sublātum, 3, *lift up, carry off, remove*
tracto, 1, *discuss, handle, manage*
trādo, -didi, -ditum, 3, *hand over*
transactor, -ōris, m. *manager, agent*
transigo, -ēgi, -actum, 3, *settle, finish*
tribūnal, -ālis, n. *platform, dais*
trīclīnium, -ii, n. *a table-couch*
triennium, -ii, n. *a period of three years*
triumpho, 1, *hold a triumph, am glad*
trulla, -ae, f. *a dipper*
tumulus, -i, m. *mound, hill*
tunica, -ae, f. *tunic*
turbulentus, adj. *boisterous*
turpitūdō, -inis, f. *disgrace, shamefulness*
tūtus, adj. *safe*

ubīque, adv. *in any place whatever*
ulciscor, ultus, 3, dep. *avenge*
ullus, adj. *any*
ultrā, adv. *beyond, besides*
umquam, adv. *ever*
ūnā, adv. *together with*
unde, adv. *whence*
undique, adv. *from all sides, everywhere*
ūniversus, adj. *whole*
ūnus quisque, adj. and pron. *each single* (one)
usque, adv. *as far as*
ūsus, -ūs, m. *use*
uterque, pron. and adj. *each of two, both*
ūtilitās, -ātis, f. *usefulness, value, importance*
ūtor, ūsus, 3, dep. *use, experience, enjoy*

valva, -ae, f. *leaf of a door, a folding door*
varius, adj. *different*
vās, vāsis, plur. **vāsa**, -ōrum, n. *vessel, utensil*
vehemens, adj. *furious*
veho, vexi, vectum, 3, *carry*
vel, conj. and adv. *or, either, certainly*
vēlum, -i, n. *awning, covering, sail*
vēnāticus, adj. *hunting*
vendo, -didi, -ditum, 3, *sell*
venio, vēni, ventum, 4, *come*
venustās, -ātis, f. *charm, grace*
venustē, adv. *charmingly*
vēr, vēris, n. *spring*
vereor, veritus, 2, dep. *fear*
vērō, adv. *indeed, however*
verso, 1, *turn about often*
versor, 1, dep. *am engaged in, live, am*
versus, prep. + acc. *towards*
verto, -ti, -sum, 3, *turn*
vērum, adv. *but, truly*
vestīgium, -ii, n. *track*
vestio, 4, *clothe*
vestis, -is, f. *clothes*
vestītus, -ūs, m. *clothes*
veto, -tui, -titum, 1, *forbid*
vetus, adj. *old*
vetustās, -ātis, f. *age, antiquity, standing*
vexo, 1, *harass, annoy*
victōria, -ae, f. *victory*
vidēlicet, adv. *evidently*

video, vīdi, vīsum, 2, *see*
videor, vīsus, 2, dep. *seem, seem good*
vigilantia, -ae, f. *watchfulness*
vigilia, -ae, f. *wakefulness, watching*
vigilo, 1, *am awake*
villa, -ae, f. *country house*
vīnārius, adj. *belonging to wine*
vincio, vinxi, vinctum, 4, *bind*
vinculum, -i, n. *a chain*
vīnum, -i, n. *wine*
violo, 1, *outrage, violate, profane*
virga, -ae, f. *rod*
virginālis, adj. *belonging to a girl*
virgō, -inis, f. *maiden*
virtūs, -ūtis, f. *goodness*
vīs, vim, vī, f. *force, violence, quantity*; (in plur.) *strength*
vīso, -si, -sum, 3, *view, visit*
vīta, -ae, f. *life*
vitium, -ii, n. *crime*
vīto, 1, *avoid*
vituperātiō, -ōnis, f. *blame*
vīvus, adj. *alive*
vix, adv. *scarcely, hardly*
voluntās, -ātis, f. *willingness, inclination*
vōx, vōcis, f. *voice*
vulgō, adv. *publicly*
vultus, -ūs, m. *expression*